Physics of Electronics for PCB Designers

(Without the Formulas)

Douglas Brooks, PhD
2022

Copyright
Douglas Brooks
Issaquah, WA
2022

Table of Contents:

		Page
00	Preface	3
0	Introduction	5
1	Current, Electromagnetic Fields and Signal Basics	7
2	Field Impacts on Components	15
3	Field Coupling Impacts	23
4	Field Radiating Impacts	31
5	Field Arcing Impacts	37
6	Physics of Trace Temperatures	41
7	Some Material Impacts	49
A1	Appendix 1: Basic Waveform Characteristics	57
A2	Definitions of Physics, Electrical Engineering and Electronics	59
	Other Books by the Author	61

The Physics of Electrical Engineering for PCB Designers

Preface:

When we design PCBs, are we involved with Physics, Electrical Engineering, or Electronics? One very legitimate response to this question might be "Who cares!" Or, "So what!" Or even, "What's the difference?" And the difference might be pretty small and arbitrary. In Appendix 2 I have offered some of the definitions of these terms you can find on the internet. They aren't very helpful.

For the purpose of this booklet I am going to lay out these broad guidelines (straight from Appendix 2):

Physics is the natural science that studies matter, its fundamental constituents, its motion and behavior through space and time, and the related entities of energy and force.

Electrical Engineering is a subset of physics dealing with electricity and electrical systems.

Electronics is a subset of Electrical Engineering that deals with electrical circuits and components.

PCB designers tend to have training (if they have training at all) in electronics. We can think of the world of electrical circuits and components as being that of resistors, capacitors, inductors, conductors, impedance, and semiconductors, for example. It is a world where Ohm's Law (E=IR or E=IZ) applies. It is a world where power dissipation is equal to I^2R. It is a world where current flows down a trace, driven by a voltage, and if we lay out the traces correctly, the circuit will perform as designed. It was a comfortable world for most PCB designers until about the mid 1990's.

What we didn't need to know until then, was that there was more to electronics than this. When frequencies (actually rise times) got high (fast) enough, we found that there were some physics phenomena associated with force fields that became pretty important. The movement of current resulted in magnetic and electric charged fields that created capacitive and inductive effects we hadn't had to deal with before. The movement of current resulted in energy (in the form of electromagnetic waves) being radiated outwards away from the circuit causing various forms of unexpected interference. *"Motion and behavior through space and time, and the related entities of "energy" and "force"* became important issues. And we designers had little background and experience in this area of physics. But this intrusion of some physics properties into our lives required us to become familiar with terms like EMI, cross talk, controlled impedance traces, and power system integrity. In some cases it even required us to worry about the length of a simple via.

The physics of materials also became an issue. We found that material properties can impact such things inductance, capacitance, and signal timing. Even something as benign as the lay of a fiberglass weave on a PCB can have an impact on trace impedance and signal timing.

And the physics of thermal (heat) conduction, convection, and radiation has a huge impact on trace and via temperature.

It is the influence of the properties of physics that:
    Causes capacitance and inductance
    Causes a circuit to work at lower frequencies but not higher frequencies
    Causes a signal to radiate to an FTC compliance antenna
    Causes the skin effect in a conductor
    That causes trace temperature to be correlated to current (sometimes) but
        via temperature to not be correlated to current at all
    Changes signal timing
    Causes thermal vias to be ineffective

The purpose of this booklet is to introduce some of the relevant physics phenomenon to board designers, and to explain WHY that phenomenon is important. Then to point the reader to areas where they can learn more about that specific phenomenon. The purpose is NOT to provide the equations and formulas for solving the problems (although there is a little of that here.) I want you to come away with an *understanding* of the physics you need to understand. To come away with an understanding of WHY you need to understand it. And to come away with the knowledge of where to go to get more details. I hope you find this booklet worthwhile.

You don't need a strong background to get a lot out of this book. But it is written for PCB designers, so we do assume some basic understanding of PCB design principles and some basic electronic background.

Further Reading:

"PCB Currents; How They Flow, How They React," Doug Brooks, Prentice Hall, 2013

"UltraCAD's Best Articles and Application Notes," Doug Brooks, 2022, available on Amazon.com

"Maxwell's Equations Without the Calculus," Douglas Brooks, 2016, available on Amazon.com

"PCB Design Guide to Via and Trace Currents and Temperatures," Brooks and Adam, Artech House, 2021.

0.0 Introduction:

This book is written for PCB Designers. It is assumed that the reader is familiar with important PCB design issues including such things as controlled impedance traces, differential impedance, cross talk, and EMI. It is also assumed that the reader has had some exposure to important electronics concepts such as Ohm's Law, resistance, reactance and impedance, and transmission line terminations.

What is unfamiliar to a great many designers is that signal propagation also creates electromagnetic energy that can have a significant effect on the design and the board's ultimate performance. This is part of where the *physics* (as opposed to the electronics) of the system comes into play.

Designers may have had some courses (or even a degree) in electronics. But they have probably had little exposure to *physics*. Physical properties impact electronics (and in particular PCB design) in at least three important ways:

1. Electromagnetic energy and its effects
2. Material properties and designs
3. Thermal conduction, convection, and radiation.

This is NOT a book on *physics*. This is NOT a book on how to *quantify* the effects of the impact of physics. In fact, there are very few equations at all in this book, and those that do appear are generally pretty simple. This book provides a *qualitative* discussion of some of the physics that impact PCB design with the purpose of providing the designer with the *What, How,* and *Why* of these issues. If you want to know *How Much* these issues impact designs, you need to read texts that go much deeper into the topics.

Probably the most important impact of physics on board design is through electromagnetic energy. Electromagnetic energy is a direct result of changing current along a trace. It is the root cause of inductance. It relates to capacitance. It causes EMI and cross talk, impacts differential impedance, and of course carries wifi signals all around your home.

But there are other physics concepts that are important, too. For example:

Materials differ in their resistivity, ρ
Materials differ in their permeability, μ_o
Materials differ in their thermal conductivity, k
Materials differ in their relative dielectric coefficients (permittivity), ε_r

Materials sometimes differ in the way they are fabricated. For example, a tight fiberglass weave in a dielectric will have different impact on the signal than will a loose fiberglass weave.

Again, this book is intended to give you a qualitative understanding of what these various impacts are, and when they might be important. But it is not intended to give you a quantitative guide of how to deal with them.

1.0 Current, Electromagnetic Field, and Signal Basics

1.1 What is Current?

If you Google "electrical current" you get over one billion (with a B!) hits. Most of them say (in different words, I didn't review every one) that current is the flow of charged particles. In electronics, the definition reduces to the flow (or movement) of electrons.

The formal definition of one amp of current is the movement of one coulomb of charge past a point (or across a surface) in one second of time. One coulomb of charge is further defined as the charge on 6.24×10^{18} electrons. The definitive measure of current would be the actual count of electrons passing across a surface. We don't know how to do that, so all measures of current *infer* that flow by some other property of current, most commonly by a voltage drop across a known resistance or by the magnetic field the current generates around the conductor as the electrons flow.

Materials (elements) that are good conductors have a single electron in the outer (valence) shell. Copper, silver, and gold are known as good conductors and they each have two very important properties in common: they are solids at room temperature and they have single electrons in their outer, or valence, shells (or "bands.") These outer electrons are not tightly held to their parent atoms. In fact, it is not always clear which atom "owns" which of these "outer" electrons. They are often (inappropriately) referred to as "free" electrons. As electrical force (voltage) is applied to a conductor, electrons tend to "jump" from one atom to the next (Figure 1.1), constituting the "flow." If you have a length of copper wire, it is not wrong to conceptually think about one electron popping out the far end for each electron that is injected somehow into the front end.

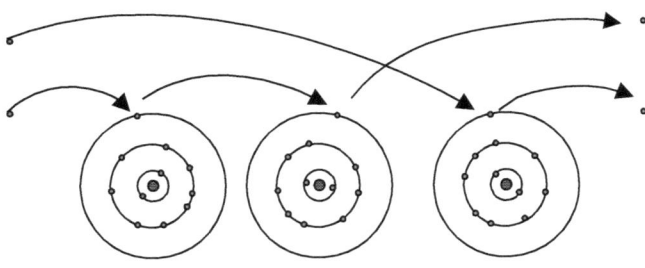

Figure 1.1
Single electrons in the outer, or valence, shell are free to move among similar atoms.

The "flow of electron" definition is not without its critics or without criticism. While it is true that current is the "flow of electrons," the term can be misleading if we are not careful in how we describe it. Some people (incorrectly) refer to this idea as the "fluid model" of current flow, sort of like water flowing through a pipe. While the analogy of water flowing through a pipe is useful for illustrating some types of current concepts (such as resistance, capacitance and transmission line reflections) it is a very poor analogy in other ways. In particular, the fluid model (correctly) and the "flow of electrons" model (incorrectly) are criticized as being unable to deal with another aspect of current --- that it propagates at the speed of light. In fact, I have heard people take the absolute position that current *is not* the flow of electrons because electrons do not flow at the speed of light.

Signal flow at the speed of light: Here is where communication breaks down. When we talk about current being the flow of electrons and current flowing at the speed of light, we are *not* talking about the same electron flows! Individual electrons actually flow very slowly through a conductor, much, much slower than at the speed of light. But electrons (as a group) *shift* very quickly through a conductor, almost, well…. at the speed of light! We have to focus on the right thing.

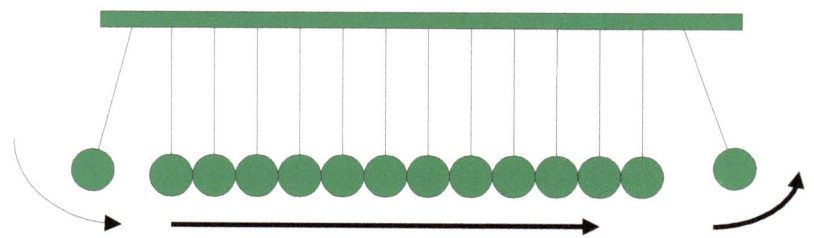

Figure 1.2
Electron flow causes a shift in electrons along a conductor.

Figure 1.2 illustrates what I mean. We are all familiar with this desktop toy. The falling ball at the front hits the first ball and transfers its energy to it, which then transfers energy to the second ball, and so on. The last ball in the string pops out (in the absence of friction) as far as the first ball fell. We could think of this string of balls as being VERY long. Even so, in the absence of friction, the last ball would pop out almost instantaneously, almost at the speed of light, even though the first ball is not moving nearly that fast. This is a rough analogy of what happens with electrons. And the distinction is very important.

Current flow does not mean "one electron in – *same* electron out." This happens very slowly as that individual electron travels through the conductor. Instead, it means "one electron in-one electron out." This can happen almost instantaneously, as Figure 1.2 suggests. (Note 1)

Summary: In summary, the "flow of electron" definition of current is perfectly appropriate as long as we understand what electron flow is. None other than the Nobel Prize-winning physicist Richard Feynman confirms the electron component of current in his book QED. (Note 2)

> … *In an atom with three protons in the nucleus exchanging photons with three electrons – a condition called a lithium atom – the third electron is further away from the nucleus than the other two (which have used up the available space), and exchanges fewer*

photons. This causes the electron to easily break away from its own nucleus under the influence of photons from other atoms. A large number of such atoms close together easily lose their individual third electrons to form a sea of electrons swimming around from atom to atom. This sea of electrons reacts to any small electrical force (photons), generating a current of electrons – I am describing lithium metal conducting electricity. Hydrogen and helium atoms do not lose their electrons to other atoms. They are "insulators."

1.2 Electromagnetic Field Basics

Charles-Augustin de Coulomb (1736-1806) was a French physicist. He is best known (at least to us) for developing a couple of laws way back in the 1700's.

There are two types of charge, positive and negative. Unlike charges attract and like charges repel each other (Figure 1.3) with a force that is proportional to the product of their charge and inversely proportional to the square of the distance between them.

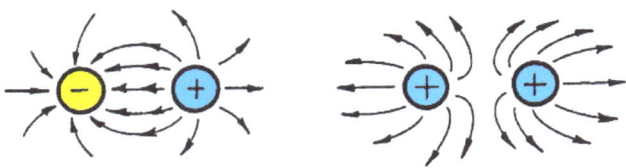

Figure 1.3
Electric lines of force illustrate Coulomb's Law

Important note about these forces: Imagine these charged particles are electrons. Then imagine we had n1 electrons at one point and n2 electrons at a second point and n3 electrons at a third point. Finally assume that n1 > n2 > n3. **The force between any two points is the voltage between them**, so in this imagined circumstance $V_{n1-n3} > V_{n1-n2} > V_{n2-n3}$. Voltage is the force between charged particles, the electric field is made up of the lines of force between the charged particles.

Coulomb also gave us another law related to magnetism: *Every magnetic pole is a dipole with an equal and opposite pole.* That is the same thing as saying that a magnetic "north" pole cannot exist without there also being a magnetic "south" pole. Even if you cut a magnet in half (see Figure 1.4), the individual poles would not be preserved; new poles would appear to preserve the dipole nature of the magnet.

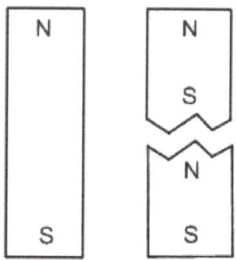

Figure 1.4
Every magnetic pole is part of a dipole.

This leads to a similar law as above:

> *Magnetic force is a vector whose direction is a line along which the force acts (Figure 1.5). This magnetic force is inversely proportional to the square of the distance.*

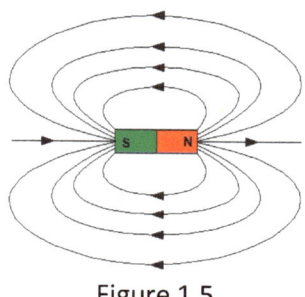

Figure 1.5
Magnetic lines of force

Lines of force, as described above, constitute a "field". So, Coulomb's two laws describe a charge (which we will interpret as "electric") field and a magnetic field, the strength of each being inversely proportional the "square of the distance," (the *square* law.)

Then, Andre-Marie Ampere (1775-1836), a French mathematician and physicist, is credited with formulating Ampere's Law in 1825 (see Figure 1.6):

> *An electric current is accompanied by a magnetic field whose direction is perpendicular to the current flow. (This is the basis for an electro-magnet.)*

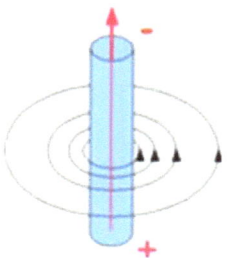

Figure 1.6
Ampere's Law

There is an extension of Ampere's Law, credited to Maxwell (Note 3):

> A **changing** electric current is accompanied by a **changing** magnetic field.

And finally, Michael Faraday (1795-1867) had little formal training as a scientist. He was what we might refer to today as a "lab rat." Most of what he discovered he did so empirically by experimenting in the lab. He is credited with developing Faraday's Law of Magnetic Induction in 1831 (Note 4).

> A **changing** magnetic field is accompanied by a **changing** electric field perpendicular to the change of the magnetic field.

Ampere's Law and Faraday's law combined together explain the principles behind motors and generators.

Important Note on Magnetic Fields: We learned in EE101 that if we send a current down a trace, a magnetic field is formed around the trace. It takes some energy to create that field. And that energy is stored in the field. But that is only a small portion of the energy flowing down the trace. We know that because the energy in the trace heats the trace by the I^2R power dissipation in the trace. Then if we stop the current flow, the energy in the magnetic field collapses back around the trace and returns to the trace. In an ideal system, *there is no energy lost in the field*. Only the "real" term in the impedance expression (the resistance) can result in energy loss (Note 5). The energy that builds up around (ideal) inductors and capacitors is always returned back to the circuit.

What It Looks Like: Figure 1.7 illustrates what the electromagnetic field would look like for a differential signal pair suspended in the air. Electrical field lines (blue) extend between them from one conductor to the other. Magnetic lines of force (red, field lines) circle around each conductor. The magnetic field would normally be circular for a single conductor (see Figure 1.6), but in this case, since there are two fields of opposite polarity, the two fields are squeezed between the conductors.

The closer the two conductors are to each other:
    (a) the shorter are the electric field lines and
    (b) the tighter are the magnetic lines are squeezed together, and therefore
    (c) the stronger is the coupling between the conductors.

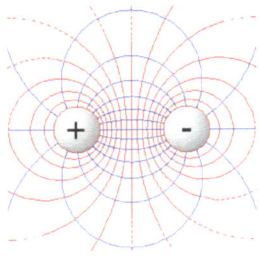

Figure 1.7
Electromagnetic field illustration

Figure 1.8 (Note 6) illustrates what the electromagnetic field lines might look like for a differential signal pair of traces in a Microstrip configuration. If the current on one trace is (electrically) positive and the other (electrically) negative, the traces will attract each other (opposite charges attract.) You can see the electric field lines going from one trace to the other and also from the trace to the return plane. Since the differential currents are equal and opposite, the magnetic fields have opposite polarities. The magnetic lines circle each trace and are "squeezed" between them, showing that there is coupling between them. But there are no magnetic field lines that will encircle *both* traces.

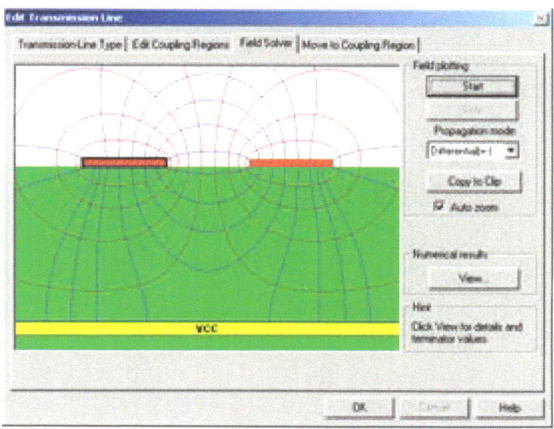

Figure 1.8
Electric and magnetic fields around a differential pair of traces.

Summary: Putting all this together results in the following:

1. Current is the movement of charged particles (electrons).
2. Electrons, being charged particles, generate electric fields whose force is subject to the square law.
3. Electrical current flowing through a wire generates a magnetic field around the wire whose force is subject to the square law.
4. If there is a current flowing in a wire, both fields are being generated at the same time.
5. Collectively these fields (electric and magnetic) are called the "electromagnetic field."
6. If the current is changing (i.e. there is an electronic signal), then the electromagnetic field is also changing.
7. These three components (electric field, magnetic field, and current) are coincident and move along together. That is, the electromagnetic field can't get in front of the current, the current can't get in front of the electric field, etc.

1.3 What is a signal:

For PCB designers, a "signal" is a changing current or a changing voltage along a trace. From Ohm's Law, we know that if there is a conductor with any resistance at all, then there will be *both* a current and a voltage (except for the trivial case where there is neither.) So, it doesn't matter if we think of the signal as a changing current or a changing voltage. If there is a changing current, then there is both a changing electric field and a changing magnetic field around the conductor. Therefore, there is a changing electromagnetic field around the trace.

1.4 WHERE is the signal?

When I first got involved in Printed Circuit Board Design (early 90's), fast signal rise/fall times were just starting to become a general issue. Prior to that we had been pretty much a "connect-the-dots" kind of discipline. But as rise times got faster, it became necessary to worry about (electromagnetic) fields. One manifestation of that was EMI (see Section 4.2), and the increasing need to pass FTC compliance testing.

So, a new type of engineer came on the scene, the Electromagnetic Compliance/Compatibility Engineer. Up until that time, we understood that (electrical) current on a copper trace (the signal) was the "flow" (movement) of electrons along the trace ($6.25*10^{18}$ electrons crossing a surface in one second of time.) But these new engineers came along and (many) started saying things like (Note 7):

> No, current isn't electron flow. Electrons can't flow at the same speed signals flow (DGB: but they CAN transfer energy between themselves at the speed of light, which is how they "flow").
> Maxwell (1867), and Maxwell's Equations tell us that the signal is in the field around the trace, not on the trace itself.
> And even...... Stop worrying about traces; ignore them. Just control the fields and you will be fine.

These engineers argued that the signal was somehow not on the trace. It existed only in the field around the trace. That has resulted in the question "Just where *IS* the signal, on the trace or in the field around the trace?"

This is NOT an either/or question. You can see from the above discussion that the signal is inherent *everywhere*. It is in the current along the trace. It is in the voltage gradients along the trace. It is in the electromagnetic field around the trace. After reading the next few chapters you will know this because:

> We can *measure* the changing current along the trace with a probe.
> We can *measure* the changing voltage gradients along the trace caused by IR drops (Ohm's Law.)
> We can *measure* the EMI radiated from around the trace.
> If we cut the trace, the electromagnetic field stops.
> If we change a characteristic of the trace (physical size or resistance), then the current *and* the electromagnetic field around the trace change.
> If we change the terminating resistor at the end of the trace, changing the reflection coefficient, the current, signal, and electromagnetic field all change.
> If we change the ε_r of the material around the trace (where the field is) the propagation speed of the (signal) current in the trace changes.

I am not arguing that the fields don't matter or are not important. They are *VERY* important and the PCB designer needs to understand them. But the current is equally important, and the designer needs to understand that, too. The remaining chapters in the book point out various relationships between the current and the fields, and how understanding both can lead to much better (and easier) design projects.

Notes:
1. It should be noted that current *must* flow in a closed loop. Thus, we cannot think about "one electron out" unless there is a path for electrons to return back to the source through that closed loop. A corollary to this is that *every signal must have a return*.
2. Richard P. Feynman, QED, The Strange Theory of Light and Matter, Princeton Scientific Library, 1985, p113
3. While *some* engineers believe this was a remarkable insight by Maxwell, others believe that it is rather obvious that if a current causes a magnetic field, a changing current would result in a changing magnetic field!
4. I have written a small booklet, available on Amazon.com, titled "Maxwell's Equations Without the Calculus." This booklet describes how these four principles are the bases for Maxwell's Equations.
   Side note: James Clerk Maxwell was a mathematician, not an engineer or physicist. His contribution to all of us was recognizing that these same laws could be combined into a "closed" system and he wrote the equations for them. An outstanding biography related to Faraday's work and then Maxwell's contribution is "Faraday, Maxwell, and the Electromagnetic Field: How Two Men Revolutionized Physics" by Nancy Forbes and Basil Mahon, 2019, available on Amazon.com.
5. The situation is slightly different in electronic transmission (as from an antenna) where we account for energy loss through a virtual *radiation resistance*.
6. See "UltraCAD's Best Articles and Applications Notes," 2022, Section 3 for significant discussions on electromagnetic fields. Available on Amazon.com.
7. I will try not to let my bias show through here. But it has been my observation that many such engineers think they are superior to all others because *they think* they understand Maxwell's Equations and you don't. Nonsense. As you will see, there is no choice of "either/or" here. And the dirty little secret is that most of *THEM* can't solve a set of Maxwell's Equations either!

2.0 Field Impacts on Components (Note 1)

2.1 Current Through Capacitors:

2.1.1 DC Current: Recall that current is the flow or movement of electrons and that electrons must flow in a closed loop (Section 1.1 and 1.2). But there is no direct electrical connection between the two parallel plates of a capacitor. The schematic of a simple capacitive circuit shown in Figure 2.1 raises the question, "How do electrons flow around the loop when there is no electrical connection between the plates of the capacitor?" The short answer is that they don't. But, of course, the total answer isn't that simple.

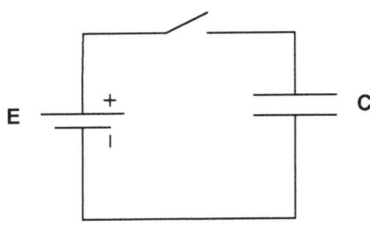

Figure 2.1
Current flow in a capacitive circuit.

Referring to Figure 2.1, assume there is no charge on the capacitor and then the switch closes, completing the circuit. At the first instant of time, electrons flow onto one plate of the capacitor. Those electrons repel (*like* charged) electrons from the other plate of the capacitor. *Those* electrons continue to flow around the loop, completing the electron flow. Recall that electron flow is really a *shift* of electrons around the circuit (refer back to Section 1.2). At the first instant of time, electron shift occurs as if there were a direct connection across the plates. The requirement of *one electron in, one electron out* is met. *This is accomplished simply by Coulomb's well-known* **physics** *fact that like charges repel each other.*

The next instant of time is a little more complex. After the first instant of time, there are more electrons on one plate of the capacitor than there are on the other plate. That is, there is a difference of charge (a voltage) across the plates of the capacitor. A difference of charge is actually the definition of voltage (See again Note 1). So, after the first instant of time, there is a voltage across the plates. That voltage tends to resist the additional flow of electrons.

So, in the second instant of time, electrons still flow onto one plate of the capacitor, and off the other, but at a slower rate. The rate is slower because a charge is building up on the plate, resisting the flow (like charges repel each other). At the third instant of time, electrons still flow, but at a still slower rate, because the voltage across the plates continues to build.

Finally, the voltage across the plates of the capacitor (the charge difference) increases to the point where it equals the driving voltage, E. At that point no further electrons can flow onto the plate. That is, no further current can flow. The process stops. All of this is a direct result of Coulomb's Law of *physics* regarding like and unlike charges.

So, in the simple circuit shown in Figure 2.1, current, or electron flow (shift), can occur until the voltage on the capacitor builds up to the same level as the driving voltage, at which point all current flow must stop. If we were then to open the switch, and if the capacitor were ideal, that voltage would remain on the plates indefinitely. (All practical capacitors have varying degrees of leakage of charge off their plates, depending on their structure and technology, so no practical capacitor can truly hold its charge "indefinitely.")

2.1.2 AC Current: Let's now change our circuit slightly. In Figure 2.2 we show a current source placed across a capacitor. Assume we let current flow in the positive direction (as shown in the figure). The voltage begins to build up on the plates of the capacitor, fighting the flow of current. Now suppose we change the direction of flow of the current to negative (opposite to that shown in the figure.) The voltage on the plates now *attracts* the current flow. At the first instant of time current will flow very freely in the opposite direction. In the next few instants of time, however, voltage will build up on the plate in the opposite polarity, again fighting against the flow of current.

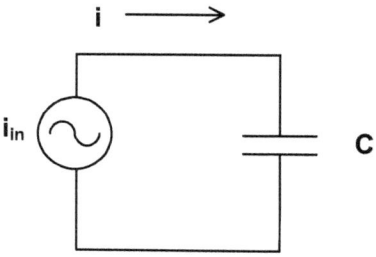

Figure 2.2
Applying AC current to a capacitor

But if we change the polarity of the current flow again, in the first instant current will again flow very freely. Each time we change the direction of the current flow, current initially flows very freely and then the current flow starts to slow as voltage builds up on the plates of the capacitor.

Consider the difference between changing the direction of the current flow very quickly (high frequency) and more slowly (lower frequency). If we change the direction of the current flow very quickly (high frequency), current flows freely. If we change the direction of the current flow more slowly (lower frequency), average current tends to be lower because more charged particles (electrons) are allowed to build up on the plates of the capacitor, repelling additional flow of the current. If we changed the direction of the current flow *very* slowly (very low or even DC current), average current becomes almost zero because enough charge builds up on the plates of the capacitor to almost completely stop the flow.

2.1.3 Ohm's Law for Capacitance: Hopefully, you are already aware of Ohm's Law for capacitance:

$$X_c = -1/\omega C = -1/2\pi f C$$

Where X_c is Capacitive reactance (impedance), Ohms
ω is the angular frequency of the signal in radians/sec
f is the cyclical frequency of the signal, in Hz
C is the capacitance in Farads
The minus sign indicates that current leads voltage with capacitance

2.2 Current Through Inductors:

2.2.1 Nature of Inductance: Recall from Chapter 1 the physical laws from Ampere and Faraday:

> a changing current causes a changing magnetic field, and
> a changing magnetic field causes a changing electric field (current)

Assume there is a very sharp change in current applied to a conductor. Suppose, for example, a signal with a very sharp rise time switches *on*. The sharp change in signal causes a sharp change in current which, in turn, causes a sharply changing magnetic field around the wire. The sharply changing magnetic field around the wire induces a current (by Faraday's Law) in the (same) wire. The direction of the induced current is *opposite* to the direction of the current that caused the changing magnetic field in the first place! Thus, the induced current fights the current generating it, or tends to cancel it out.

At the very first instant of time, the two currents are exactly equal and cancel and no current flows.

At the very next instant of time, the induced current weakens slightly (the rate of change of the changing magnetic field slows slightly) and there is a very small component of current flowing in the forward direction. In the very next instant of time a little bit more current flows in the forward direction. After some period of time, the magnetic field is no longer changing, there is no longer an induced current trying to flow in the opposite direction, and the full current flows in the forward direction.

Consider what happens when our signal switches on, is allowed to stabilize, and is then switched off. When the signal stabilizes, a full current flows in the forward direction. There is a magnetic field around the wire. Then when the current stops flowing, the magnetic field starts to collapse (that is, it starts to change.) The collapsing (changing) magnetic field induces a current, *still in the forward direction*, in order to resist the change in current that caused the collapsing magnetic field in the first place.

This is the nature of inductance. A changing current induces a changing magnetic field that acts to resist the change that is causing the flow of current in the first place. An inductor *resists* a change in current flow.

The magnitude of inductance is related to the strength of the magnetic field that is created. A magnetic field will exist around a wire if a current flows through the wire. The magnetic field can be intensified (the inductance can be increased) if any of several things occur (note that 3 and 4 depend on the *physical* properties of the materials.):

1. The wire is coiled, "focusing" the magnetic field
2. The wire is made smaller (a great big wire or plane has smaller inductance)
3. The wire is coiled around a ferrite (magnetic) material (See Section 7.2)
4. The wire is run through a ferrite bead

2.2.2 DC Current Through an Inductor: Consider the circuit shown in Figure 2.3. During the first instant when the switch is first turned on, no current flows through the inductor. In the next instant, a very small current flows. But after a few instants of time, the magnetic field reaches stabilization and stops changing. At that point there is no induced current trying to flow in the opposite direction, resisting the action of the switched current. So, after a few instants of time, a full DC current is able to flow.

Thus, we often say that an inductor offers no impedance to a DC signal (after a few instants of time). DC current can flow freely through an inductance.

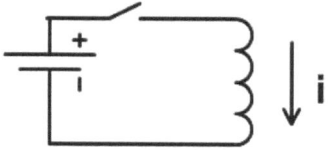

Figure 2.3
DC Current flow in an inductor.

2.2.3 AC Current Through an Inductor: Suppose, on the other hand, we change the polarity of the driving voltage before the magnetic field fully builds around the inductance. When we reverse the polarity of voltage, the magnetic field suddenly tries to reverse polarity. This causes a rapid change in the magnetic field. The changing magnetic field induces a current in the opposite direction that fights the change in driven current. If the current changes quickly again, the magnetic field tries to switch again, and so on.

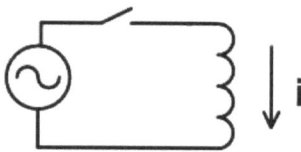

Figure 2.4
AC current through an inductor.

If the driving voltage switches polarity very fast, the current flow almost never has enough time to overcome the inductance's reactance to current change. A relatively small current is allowed to flow. Thus, the inductance looks like a very high impedance (or has a very high reactance) for very fast switching (high frequency) signals. But if the signal reverses very slowly (low frequency), then the reverse current caused by the changing magnetic field is partly overcome, and some current does flow in the driving direction. That is, there is less impedance to the current flow.

Therefore, as a rule, there is more impedance caused by an inductor to a signal if (a) the signal frequency is higher, or (b) the magnitude of the inductance (L) is higher. Inductors "impede" higher frequency signals more than they impede lower frequency signals, and larger inductors do so more than smaller inductors.

2.2.4 Ohm's Law for Inductance: Hopefully, you are already aware of Ohm's Law for inductance:

$$X_L = \omega L = 2\pi f L$$

Where X_L is Inductive reactance (impedance), Ohms
ω is the angular frequency of the signal in radians/sec
f is the cyclical frequency of the signal in Hz
L is the inductance in Hy
The (implied) plus sign means that voltage leads current (current lags voltage) through an inductor.

2.3 Skin Effect: (Note 2)

2.3.1 Inductance Causes Skin Effect:

We know that the skin effect causes the resistance of a conductor to increase with frequency. But this statement is not exactly true. Resistance is not changing. The resistance of a conductor is given by Equation 1:

$$R = \rho * L / A \qquad [Eq. 1]$$

where: R= the resistance of the conductor in Ohms
ρ = the resistivity of the conductor material
L = the length of the conductor, and
A = cross-sectional area of the conductor.

Looking at that formula, what is changing with frequency? The resistivity is not changing. The physical cross-sectional area of the conductor is not changing. The length is not changing. So why does the resistance, in fact, change with frequency?

The answer comes from Ampere and Faraday and the effect covered in Section 2.2.3 above. After a sudden change in current, the changing magnetic field causes a "reverse" current in the conductor. The answer to what is changing is *where* that current is flowing. At very high frequencies the induced reverse current flows where the induced magnetic field is the strongest. That would be back along the centerline of the conductor. The forward current tends to flow along the outer circumference of the conductor, where the induced magnetic field is weakest. Therefore, the *effective* cross-sectional area of the conductor is changing. Therefore, we have a *higher* current density through a *smaller* portion of the conductor, resulting in a higher voltage drop. This *behaves as if the resistance* of the conductor itself is changing. Skin effect is all about current density --- *where* the current is flowing.

2.3.2 WHERE Does Current Flow?: Intuitively we can picture three possibilities; (a) the net current flows uniformly across the entire cross-sectional area of the conductor, (b) the net current flows down the centerline of the conductor, or (c) the net current flows on the outside circumference of the conductor. The determining factor is the strength of the changing magnetic field. It turns out, the induced magnetic field is strongest along the centerline of the conductor and weakens inversely with the square of the distance from the centerline. Therefore, the changing field is weakest around the circumference.

Therefore, when the first increment of current starts to flow, it flows in a thin band around the circumference. As more net current flows, the band around the circumference begins to thicken. The current density is highest around the circumference, but lower current densities exist as we look further into the conductor. Only when the magnetic field stops changing (DC) does the current flow uniformly across the entire cross section of the conductor. Figure 2.5 illustrates this relationship.

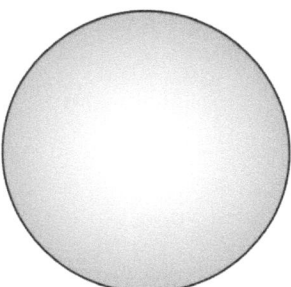

Figure 2.5
Very high frequency currents flow primarily at the circumference of a copper conductor, an effect called the skin effect. The shading represents the current density.

There are two, mutually exclusive, viewpoints regarding where the current flows:

1. Current flows uniformly down to a "skin depth" and then there is no current flow below that.
2. The current density decays exponentially from the surface of the conductor down to the centerline of the conductor.

The second viewpoint represents "truth." But we make calculations based on the first because that viewpoint and those calculations make things significantly easier. Calculations based on the difference between these viewpoints rarely vary by more than 2%, so the simplification is considered reasonable, particularly since there are other, unaccounted for, variables that also exist.

2.3.3 Why Do We Care? The first reason we might care about the skin effect is that it impacts any calculation that involves Ohm's Law (E = iR). Thus, the voltage drop across a conductor and the power dissipated within the conductor (and therefore the energy loss in the circuit) will increase with frequency.

This leads directly to the second reason; the skin effect can impact trace current/temperature effects. Although it is rare for large power supply currents to operate at very high frequencies, it is not impossible. If that happens, then the skin effect needs to be considered.

Finally, our models for transmission lines are typically lossless ones. Under that condition, the characteristic impedance, Zo, is resistive and terminating the trace is straightforward:

$$R_L = Z_0 = \sqrt{\frac{L}{C}}$$

But if there are skin effect losses (or dielectric losses, see Section 7.1), the transmission line no longer meets the conditions for being "ideal" and the termination issue becomes much more difficult (see again Note 2):

$$Z_0 = \sqrt{\frac{R_s + j\omega L}{G + j\omega C}}$$

Where:
    L is the characteristic inductance of the transmission line.
    C is the characteristic capacitance of the transmission line
    ω is the angular frequency
    J is the imaginary operator, $j = \sqrt{-1}$

    R_s is the impact of the skin effect
    G is the impact of the dielectric losses (see Section 7.1)

Notes:
    1. For a good, thorough discussion on capacitance and inductance, see Brooks, Douglas, "PCB Currents; How They Flow, How They React," Prentice Hall, 2013, Chapters 5-7.
    2. For a good discussion of skin effect, read "UltraCAD's Best Articles and Applications Notes," Amazon.com, Chapters 14-17

3.0 Field Coupling Impacts

3.1 Coupling to an Adjacent Conductor or Plane

We have learned that a changing current (electronics) through a conductor creates a changing electromagnetic field (physics) around the conductor. We have also learned that a changing electromagnetic field (physics) can induce a changing electric field. If there is metal nearby, that induced changing electric field will induce a changing current (electronics) in that metal. So, if we have a conductor (A) carrying a changing current next to a second conductor (B), a current will be induced in conductor (B). That current will flow in the opposite direction to the current that induced it.

One of the most fundamental rules in electronics is that current must flow in a loop. If electrons enter a conductor at one end, electrons **must** leave the conductor at the other end. So, if a driver injects a signal into one end of a trace, there *must* be a return signal back to the driver. *Every signal must have a return* (Note 1).

Consider a situation where the outgoing signal is on a conductor and the return signal is on another conductor (Figure 3.1). The outgoing signal on the conductor generates an electromagnetic wave which couples onto the return conductor (with coupling coefficient, k), reinforcing the return signal. The return signal also generates and electromagnetic wave, and it will couple back into the outgoing conductor, reinforcing the outgoing signal. Since the system is symmetrical, the return signal couples into the primary signal with the same coupling coefficient, k.

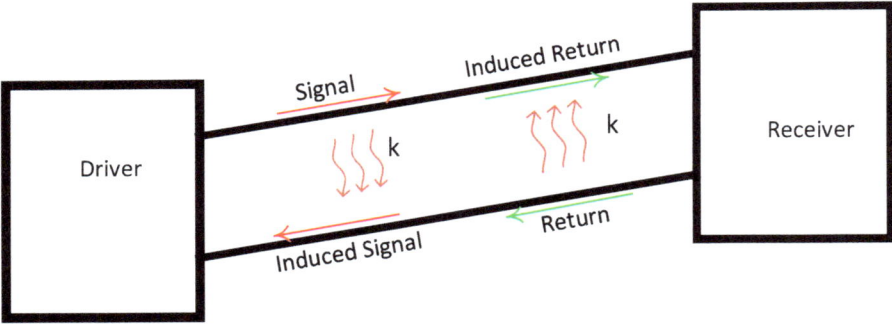

Figure 3.1
An outgoing signal and its return.

Note that these signals (outgoing and return) reinforce each other. That reinforcement lowers the "energy" the outgoing signal must have to complete the circuit. If the return conductor is moved closer, the coupling is stronger. That means even less energy is required. In the limit, if the two conductors were coincident, the coupling would be 100 percent and no energy would be required, (except that in that limit the system would be shorted out!) If less energy is required, that is the same as saying the impedance to the signal flow has decreased. In textbooks this is referred to as *mutual inductance*, so the mutual inductance decreases as the conductors move closer together (Note 2).

Now, consider a Microstrip trace over a plane (Figure 3.2 illustrates two such traces). There is a signal flowing down the trace and a return signal back on the plane. That return signal could be anywhere on the copper plane. But the *physics* of electromagnetic fields and coupling determines precisely where that return signal will be. The outgoing signal on the trace couples onto the plane and will reinforce the return signal. The return signal also generates and electromagnetic wave, and it will couple back into the trace, reinforcing the outgoing signal. From the above, it is clear that the mutual inductance will be lowest, (i.e. the impedance will be lowest, i.e. the total energy will be lowest,) if the return signal is directly underneath the trace carrying the outgoing signal. That is why the return signal will always be directly underneath the trace (if possible).

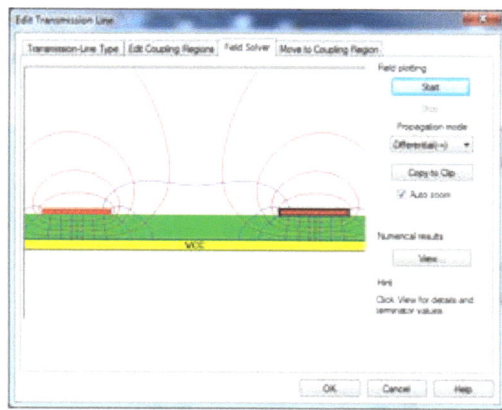

Figure 3.2
The return signal will be directly underneath the trace, where the least amount of total energy will be required.

3.1.1 Differential Impedance (Note 3): In the illustration above, I described two traces with the signal and its return (Figure 3.1). I pointed out that each signal couples into, and reinforces, the other. The stronger the coupling (i.e. the closer the traces are to each other) the stronger the reinforcement and the lower the energy required to send the signal down the trace and back.

This same illustration could apply to a differential pair. In a differential pair example we would have a positive signal on one trace (A) and its (equal and) opposite signal on Trace (B). (The returns could be on an underlying plane, but in a perfect world the returns would perfectly cancel.)

Let's let (A) and (B) be controlled impedance traces with impedance Z_0. The total impedance (down and back) would be 2 x Z_0, since the traces are in series with each other. But if the traces are close together, the electromagnetic coupling between them would mutually reinforce each signal and reduce the total energy required for driving the total signal. That actually means the total impedance is lower. So the total impedance is something less that 2 x Z_0, or it is 2 x Z_0 – (coupling effect). In fact, you should be familiar with the expression

$$Zdiff = 2 * Zo * (1 - k)$$

Where k is the coupling coefficient. This is exactly the same coupling coefficient we have been talking about in this section and it is the direct result of the electromagnetic field coupling. Furthermore, you now should be able to see why if the traces are closer together, the differential impedance gets lower.

3.2 Crosstalk Coupling

Let's modify our second example above slightly. Let's assume that the signals on the two conductors are independent (and their returns are on the underlying plane.) Conductor (A) will couple into Conductor (B), and vice versa. The coupled signal from (A) to (B) is crosstalk. Note that any signal on (B) will also couple back into (A) (for the reasons discussed above in Section 3.1).

You should now know how to reduce the size of the crosstalk signal... Reduce the coupling. We can reduce the coupling by moving the traces further apart (by the inverse square law.) Or, we can reduce the coupling between (A) and (B) by moving (A) closer to the plane (Figure 3.3). That captures more of the electromagnetic field between (A) and the plane so less of the field is available to couple into (B). This is entirely consistent with what you already know. To reduce crosstalk coupling:

1. Move the traces further apart, and/or
2. Move the traces closer to the return plane.

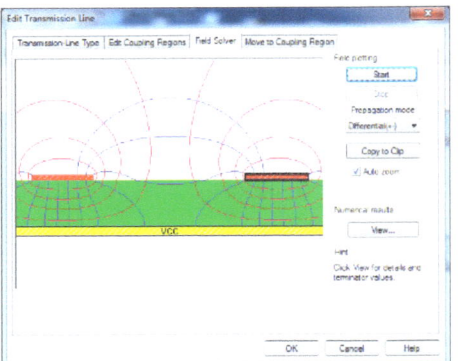
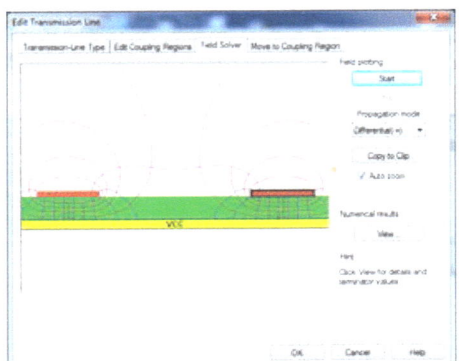

Figure 3.3
Crosstalk coupling

At this point you should be recognizing the importance of being able to *visualize* the electromagnetic field around your traces. The solutions to many signal integrity problems involves controlling the fields around the traces. By visualizing the field in your mind, you can intuitively see what it is you need to do.

3.3 Transformer Coupling:

Figure 3.4(a) illustrates two coils placed next to each other. This is just a crosstalk illustration with the adjacent conductors coiled instead of being straight. The coils provide a stronger electromagnetic field coupling between the two conductors, so we would expect there to be stronger coupling between the two conductors. In fact the coupling *is*

25

stronger, and is related to the number of turns in the coils. The more turns there are, the greater the coupling. Also, the closer the two coils are to each other, the stronger the coupling.

Now consider Figure 3.4(b). In this instance we have placed a magnetic material (core) between the two coils. The magnetic core concentrates the magnetic field and significantly increases the coupling (see Section 7.2 and Note 4). Coupling coefficients (k) approaching 99% are possible when we introduce an iron core coupling the windings.

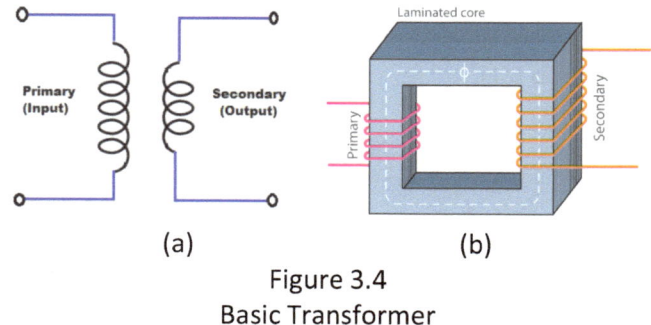

Figure 3.4
Basic Transformer

Note that we have gone from a little coupling, Section 3.1, to almost perfect coupling, Section 3.3, without talking about signal levels or trace sizes at all. All these coupling effects are the result of the *physics* of physical interactions.

3.4 Fields and Impedance (Note 5):

Figure 3.5 illustrates the electromagnetic field around a pair of 8 mil wide, 0.5 Oz. traces, 6 mils above the underlying plane. The traces form a differential pair. The trace separation is 20 mils. The single-ended characteristic impedance of each trace is 59.6 Ohms. (This results in a Zdiff of 114.9 Ohms.)

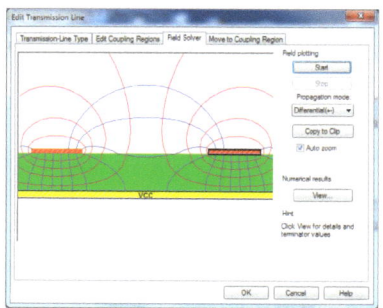

Figure 3.5
8 mil trace, 20 mil separation, 6 mil above plane, 0.5 oz thick

Figure 3.6 shows the field patterns for a pair of 16 mil traces, 12 mils above the plane, and 1.0 Oz. in thickness, separated by 40 mils. It looks a lot like Figure 3.5. In fact, it almost looks like I have repeated Figure 3.5 by mistake. But I haven't! The field patterns look identical. In fact, the field pattern **are** identical! And the single-ended characteristic impedance is identical, 59.6 Ohms. (And the Zdiff is also identical at 114.9 Ohms.) In fact, **everything is identical, just scaled.**

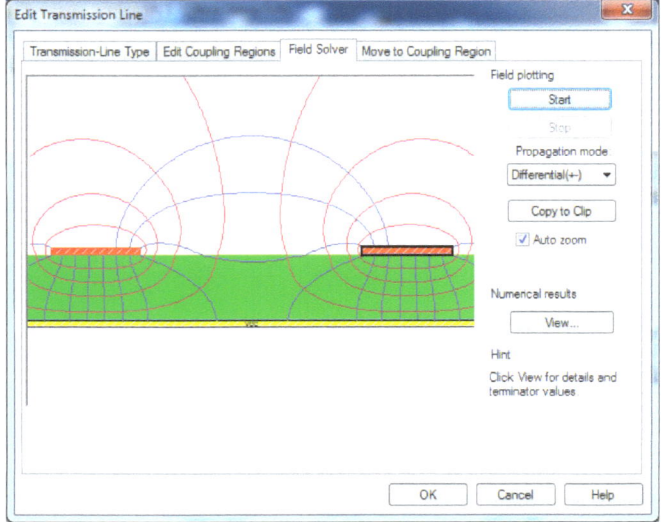

Figure 3.6
16 mil trace, 40 mil separation, 12 mil above plane, 1.0 Oz thick

This illustrates a very important point. The impedance is not necessarily determined by the stackup and trace dimensions. It is, but there are several different combinations of stackup dimensions that will lead to the *same* characteristic impedance. **The impedance of the trace is determined by the electromagnetic field distribution.** If we *scale* all the dimensions, we don't change the electromagnetic field pattern, and therefore we don't change the impedance.

By now you should be able to look at a field diagram and judge what happens to the trace impedance if we make some small changes in the trace itself. For example, what happens if:

1. We lower (decrease) the height of the trace above the plane? (The capacitance of the trace increases and the overall impedance goes down.)
2. We increase the trace width? (The capacitance between trace and plane increases and the trace impedance goes down.)

3.5 Fields and Signal Timing:

Recall that electronic signals travel at the speed of light, 186,282 miles per second. This equates to 11.8 inches/ns (or what we sometimes round off to a foot a nanosecond.) In any other material the speed of light slows down. It slows down by the square root of the relative dielectric coefficient, Equation 1.

$$Speed = \frac{11.8}{\sqrt{\varepsilon_r}} \, in/ns$$

[Eq. 1]

Consider the situation shown in Figure 3.7, derived from a Hyperlynx simulation. Here we have a trace in a Stripline environment, surrounded by a dielectric. If we assume the relative dielectric coefficient of the dielectric is 4.0, then the propagation speed of the signal will be 11.8/2 = 5.9 in/ns (we sometimes round this off to 6"/ns.) Note the electromagnetic field in this figure. It is completely contained within the dielectric between the two planes on either side of the trace.

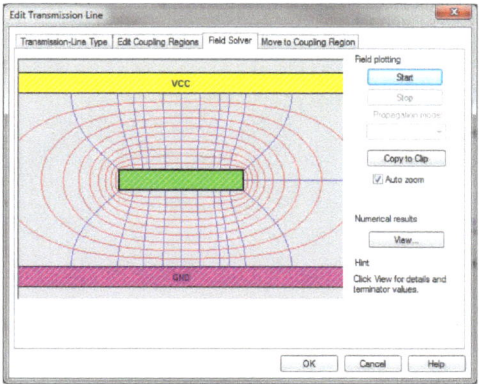

Figure 3.7
8 mil trace spaced 5 mils from each plane.

Now consider Figure 3.8. It shows the same 8 mil wide trace separated from a plane by 5 mils in a Microstrip environment. By simply looking at the electromagnetic field distribution we can make a couple of observations. Part of the field is in the same environment as in the Stripline situation, so that part will want to travel at the same speed as above (5.9 in/ns). But part of the field is in the air. It will want to travel at the speed of light in the air, 11.8 in/ns. The fields cannot travel at different speeds than the current. The electric field cannot travel at a different speed than the magnetic field. They all have to travel together at the same speed. Calculating the propagation speed is an admittedly complicated problem. But we can intuitively guess that the speed will be somewhere in between the Stripline case and pure air.

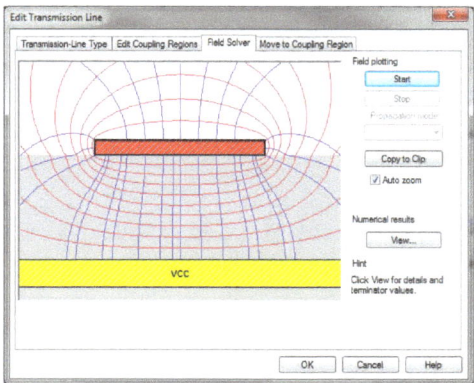

Figure 3.8
8 mil trace, 5 mils from a plane in a Microstrip environment.

This illustrates an important point:

Propagation speed does not depend on how fast the current can flow through a conductor. It depends on how fast the electromagnetic field can flow through the material it is flowing through.

Now, to illustrate this final point with more emphasis, consider Figure 3.9. It shows a 30 mil wide trace, 5 mils from a plane in a Microstrip situation. This is the same environment as in Figure 3.8, but with a much wider trace. In this situation, a significantly greater portion of the electromagnetic field is captured between the trace and the plane, with a much smaller portion of the field in the air. We can intuitively infer that the signal will slow down, compared to that in Figure 8, because more of the electromagnetic field is contained within the slower dielectric. In the limit, for an infinitely wide trace, all the electromagnetic field will be in the slower dielectric and the propagation speed will be the same as in the Stripline situation.

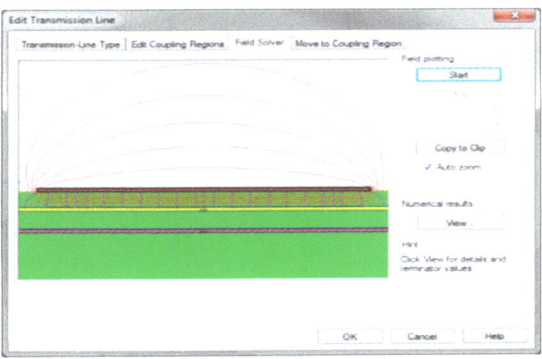

Figure 3.9
30 mil wide trace, 5 mils from a plane.

Table 3.1 shows what happens to propagation speed as the width increases in the example above. In every case, distance to any plane is 5 mils, relative dielectric coefficient of the dielectric is 4.0, and trace thickness is 0.5 Oz.

Width of trace (mil), Microstrip	Propagation Speed (in/ns)
Stripline (for reference)	5.9
8	6.9
30	6.5
100	6.2
500	6.0

Table 3.1
Propagation speed as a function of microstrip trace width.

The trend for the speed in Microstrip to approach the speed in Stripline as the trace width increases is clear.

Notes:
1. See "UltraCAD's Best Articles and Application Notes," especially Chapter 22.
2. We could introduce the formulas for mutual inductance and prove mathematically that this is all true. But if we did, we would all be so confused that we would miss the overall points. That is why this book has very few formulas!
3. See "UltraCAD's Best Articles and Application Notes," Chapter 2.
4. This illustration shows the magnetic material in the shape of a square loop. But the shape is not important, it could be a round toroid or simply a straight rod without changing the situation much.
5. This illustration was done with the Hyperlynx Simulation tool.

4.0 Field Radiating Impacts:

Broadly speaking, there are two types of electromagnetic radiation from a system, (a) intentional and (b) unintentional. Again, oversimplifying a little, intentional is good and unintentional is bad! This chapter talks about a variety of ways our boards may radiate electromagnetic energy (physics) in ways that hurt our signals (electronics).

4.1 Transmission:

Intentional radiation is transmission. We transmit a signal during a radio or tv broadcast, when we lock/unlock our car with a key fob, when we use a tv remote, and when our routers broadcast a wifi signal around our home or office. GPS satellites radiate electromagnetic energy. The objective, often, is to maximize signal strength (although some regulations require that there may be limits to signal strength.)

One way to optimize signal strength is to (impedance) match the output circuit of our transmitter circuit to the input impedance of the antenna (more on this below.) This topic is beyond the scope of this book.

4.2 EMI:

Almost all other electromagnetic radiation from our board is unintentional and harmful. There are often specific limits of radiation we must adhere to and many systems must pass FCC compliance testing before they are marketed. Radiation from one board can impact the performance of another board in the same system.

We have seen in the previous chapter that electromagnetic coupling can occur to adjacent traces on the board (crosstalk.) Similarly, the same type of coupling can occur to conductors (or other objects) off the board. When it occurs on the board, we call it crosstalk. When it occurs off the board, we call it EMI (Electromagnetic Interference.) The cause and the effect are exactly the same. You may have seen cases where a system failed FCC compliance testing and then the engineers spent significant time trying to seal various chassis openings with copper tape to try to keep the electromagnetic energy from leaking out. **This is the wrong strategy!** The proper strategy is to prevent the EMI at its source, not to cover it up somewhere else. That is what this section discusses.

4.2.1 Loop Areas: One of the absolutely fundamental truths in electronics is that current flows in a closed loop. Current is the flow of electrons, and if it were not true that current flows in a closed loop, then electrons would start collecting in some sort of pool somewhere along a wire or trace. Intuitively, we know this doesn't happen. And if current does flow in a closed loop, then it is also absolutely, fundamentally true that every signal has an equal and opposite return signal associated with it.

When we design a PC board, we carefully design the path that the signal takes. But we often don't consider the path of the return signal. We simply take for granted that the return signal will sort of take care of itself. It turns out that in high-speed designs it's pretty important for the designer to know where the return path is for each (and every) signal. It *does* exist. The only question is "Where is it?"

If current flows in a closed loop, then we can visualize the *physical* area defined by that loop. Take, for example, a twisted pair of wires with the signal on one wire and the return on the other. Since the wires are twisted closely together, the loop area is pretty small. A coax cable with the signal on the center conductor and the return on the shield also has a very small loop area. But, if we had, for example, a ten inch long trace with a return trace one inch away, we would have a loop area of 10 in^2, *much* larger than in the other two cases.

"So what?" you ask. Well, there are several possible sources of EMI on a board, but a significant one is the loop area around which a very high-speed signal propagates. *EMI is directly related to loop area*. In the case of twisted pairs and coax cables, loop areas are small and these configurations perform well from an EMI standpoint. But in the case of the signal and return traces being separated, the loop area might become significant, and such a configuration might radiate badly. That's why it is important for a designer to know where the return signal path is, and to make sure that the loop area defined by the signal and its return is as small as possible.

We almost always use power and ground planes in high-speed designs. There are a variety of reasons for doing so. One of them is that if a signal trace exists above a (power or ground) plane, the return signal will *want to* be on the plane directly below the trace. The reason for this is complicated, and was alluded to in Section 3.1. If the return signal is directly underneath the trace, the mutual inductance is lowest and the impedance is lowest. Note that a signal trace whose return is on a plane directly underneath it has a small loop area.

Well designed boards, those with planes where return signals can travel directly under their corresponding signal traces, perform well in EMI critical environments. We get into trouble when we cause the signal return to move away from under the signal trace, creating a loop. But, we usually don't do this on purpose! The rest of this section will illustrate some common design problems that cause loop areas to increase.

4.2.2: Excessive Pin Clearance: Figure 4.1 (a) illustrates a trace leading to a pin on a connector. Clearance pads are so large that there is no copper for the return trace to find its way through to a ground pin. Thus, the return signal must circle around the connector to the ground pin, causing what might be a significant loop resulting in unacceptable EMI radiation. A better strategy is to limit the clearance so there are copper paths between the pins for the return signals to follow, as shown in (b). But the best strategy is (c), making appropriate pin assignments so that there is a ground (signal return) pin near every signal pin.

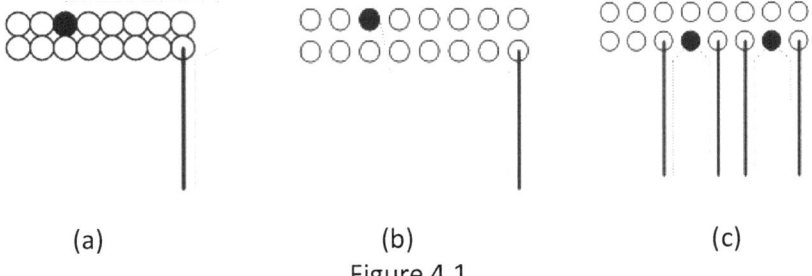

(a)   (b)   (c)

Figure 4.1
Excessive through hole pin clearance or poor pin assignment strategies can lead to excessive loop area.

Now, what if the signal trace is referenced to (is directly adjacent to) a *power* plane instead of a ground plane? The difference between a power plane and a ground plane is primarily a DC distinction. AC signals can travel with ease on any plane. So, if a signal is referenced to a power plane, how does the return signal get from the power plane to the ground pin at the connector? The practical answer is that there are usually enough bypass capacitors (between power and ground) nearby to provide a suitable path for the return signal. Some experts, however, actually recommend the placement of one or more bypass caps near a connector specifically to provide for signal return paths. (Remember, from Section 2.1.2, capacitors offer a very low impedance to high-speed signals.

4.2.3 Vias: Figure 4.2 illustrates the case of a signal moving from one signal layer to another through a via. It should be clear that the designer needs to be sure that the characteristic impedance of the trace (Zo) is the same along all segments, otherwise reflections will be caused at the vias. But what about the return signal? If the return signal has to find a circuitous loop between the various planes that are adjacent to the trace, then unacceptable loop areas (and EMI radiation) might result.

Many experts feel that while it is acceptable practice to move a signal through a via to opposite sides of the *same* plane, great care should be taken when moving a signal to a layer where it will reference to a different plane. Some experts, however, have no problem with this practice, and still others recommend placing a bypass cap near each via for the specific purpose of providing a path for the signal return.

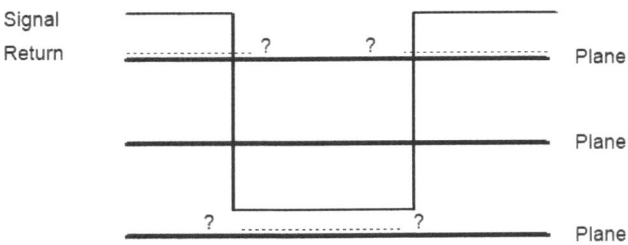

Figure 4.2
When a signal trace transitions to a different layer, it is not clear what happens to the return signal.

4.2.4 Slots in Planes: There are many reasons to avoid slots in planes. Figure 4.3 illustrates one of them. If a signal trace crosses a slot, where does the return signal go? It must find its way around the slot and a loop is inevitable. There is simply no good purpose for a slot in a plane in high-speed designs, and lots of really good reasons not to allow them.

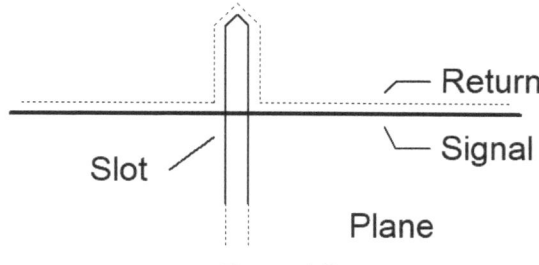

Figure 4.3
Slots in planes almost always cause loop areas to increase.

4.2.5 Crossing Unrelated Planes: We often try to isolate certain types of circuits from other ones. Separating analog circuits from digital ones is routine. An engineer might want to set up two different digital areas if it is critical that they be isolated from each other for noise purposes. Standard practice is to never allow a trace to cross over an unrelated plane. Figure 4.4 illustrates why. A digital signal trace crosses over part of an analog plane. Where will the return signal be? There are two possibilities, both of them bad! One possibility is that the return signal will find its way onto the analog plane. This may reduce the loop area but it will in all likelihood allow noise coupling between the digital and analog signals, defeating the whole purpose for separate circuits in the first place. The other possibility is that the return signal will stay on the (in this case) digital plane, resulting in a loop that might well radiate and cause EMI problems. The solution is to never route a signal over an unrelated plane.

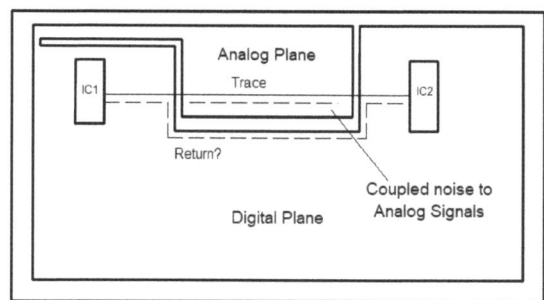

Figure 4.4
If a high-speed trace is routed over an unrelated plane, the result will be increased noise, increased loop area, or both.

4.3 Radiation Resistance:

An ideal antenna is a purely reactive device (LC circuit, no resistance.) When driven by an ideal, matched circuit, the maximum power transfer is 50% (Note 1). That means

50% of the power is transmitted, 50% of the power is "lost" to the circuit, primarily in heat dissipation. (This description also applies exactly to a driver circuit for a loudspeaker system.)

Figure 4.5 shows a simplified transmitter circuit. L1 and C2 form a resonant circuit whose impedance would match the characteristic impedance of the antenna. Now an interesting question would be, where is the 50% power loss in this schematic diagram? That is not a simple question. (It is spread between Q1 and the power supply system.) But we *could* represent it with a virtual resistor (Note 2) at the antenna input. That would suggest 50% power loss through radiation from the antenna and 50% power loss across the virtual resistor (I^2R, where R is the radiation resistance). The same explanation applies to a loudspeaker system and explains why there are 4, 8, and 16 Ohm outputs on an audio amplifier.

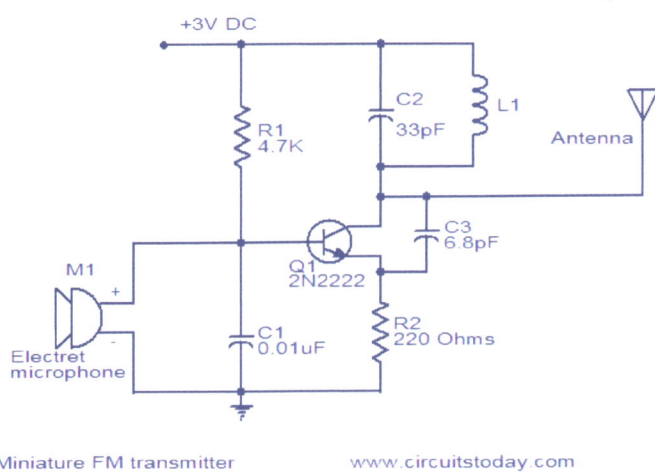

Figure 4.5
Simple FM transmitter

Looking back at EMI……
    a. If EMI is energy radiated away from our board, doesn't it represent energy lost to the system? (*YES*).
    b. Therefore, is there an antenna that is radiating it? (*Maybe. A physical loop area IS an antenna.*)
    c. And does that therefore mean there should be a radiation resistance representing this energy loss? (Yes. And *if the loss is occurring at a narrowly defined point, this makes more intuitive sense than if the loss is spread over a larger area.*)
    d. Then shouldn't we be able to measure that radiation resistance and therefore measure the actual amount of EMI radiated? (*In theory, yes. But I expect the amount is so small compared to the total energy in the system that it would be lost in the noise and not measurable.*)

Notes:

1. https://info.triadmagnetics.com/blog/maximum-power-theorem.
2. Radiation resistance is an *effective* resistance, due to the power carried away from the antenna as radio waves. Unlike conventional resistance or "Ohmic resistance", radiation resistance is *not* due to the opposition to current (resistivity) of the imperfect conducting materials the antenna is made of. The radiation resistance (R_{rad}) is conventionally defined as the value of loss resistance that *would* dissipate the same amount of power as heat, as is dissipated by the radio waves emitted from the antenna, when fed at a minimum-voltage / maximum-current point ("voltage node"). From Joule's law, it is equal to the total power Prad radiated as radio waves by the antenna, divided by the square of the rms current I_{RMS} into the antenna terminals: Rrad = Prad I^2/$_{RMS}$
    Source: https://en.wikipedia.org/wiki/Radiation_resistance

5. Field Arcing Impacts

5.1 Near Field vs Far Field Radiation:

The distinction between near field and far field radiation is not as complicated as some people make out. Consider Ampere's (as modified) and Faraday's Laws (Section 1.2):

> **Ampere:** A *changing* electric current is accompanied by a *changing* magnetic field.
> **Faraday**: A *changing* magnetic field is accompanied by a *changing* electric field at right angles to the change of the magnetic field.

This is as close a description of perpetual motion we are likely to get. Each type of energy reinforces the other, and in the absence of losses, this mutual reinforcement can go on forever. Astronomers search the universe for electromagnetic signals tracing back to the "Big Bang," 14 billion years ago. Far Field radiation refers to any electromagnetic radiation where this mutual reinforcement is occurring (a steady state condition). Any electromagnetic radiation will quickly "fall into" the far field mode no matter how it initially starts. "Near Field" radiation refers to any electromagnetic radiation that occurs between the initiation of the radiation and when far field radiation takes over (a transient condition.) Near field radiation can be initiated by any number of events. Radiation from an antenna is an obvious one. A spark jump can initiate near field radiation (Note 1), as can a rotating magnet.

An interesting question is … how long (or over what distance) does it take for energy to convert from near field to far field mode? The answer, of course, is it depends on the source of the initial radiation. If the initial radiation is from a well-designed antenna circuit, there is almost no "near field" radiation. That is, the initial radiation from the antenna is already in far field (steady state) mode. But a spark starts out almost totally as electric field radiation and far field mode needs to evolve. In practical terms, no matter how an electromagnetic field starts, it converts to far field mode within about one wavelength.

5.2: The Physics of Spark vs Arc:

What is the difference between a spark and an arc? They each start the same way. A relatively strong electric field is built up across two points. The strong electric field is created by a charge difference (voltage) between the two points (remember Coulomb, Note 2). There are one or more dielectrics (Note 3) between the two points. If the voltage (electrical force) between the points exceeds the breakdown voltage for the dielectric, charges (current) will jump from one point to the other.

Materials differ greatly in their dielectric breakdown voltages. Table 1 (Note 4) provides voltage breakdown estimates for several different materials. These estimates depend greatly on conditions, the width of the gap, temperature, material purity, etc., and should be taken as rough estimates only.

Table 5.1
Approximate breakdown voltages for various materials.

Material*	Dielectric strength (kV/inch, or V/mil)
Vacuum	20
Air	20 to 75
Nitrogen	23 to 86
Hydrogen	13 to 56
Neon	.4 to 2
Argon	4 to 20
Porcelain	40 to 200
Paraffin Wax	200 to 300
Transformer Oil	400
Bakelite	300 to 550
Rubber	450 to 700
Shellac	900
Paper	1250
Teflon	1500
Glass	2000 to 3000
Mica	5000

= Materials listed are specially prepared for electrical use.

What happens next depends on the material:

(a) A *spark* jumps when the electric field gets high enough. But it is typically a singular, discrete event. The electric field needs to be maintained to continue a current flow.
(b) An *arc* starts when the threshold (breakdown) voltage is reached, but then the current ionizes, say, a conducting medium (gas, e.g.) so that the arc can be maintained at a lower driving voltage.

A spark is a singular, transient event, and current typically flows in a short spike. An arc overcomes a threshold voltage and then current will continue under the force of a lower voltage. A gas discharge lamp (e.g. neon bulb) is an example of an arc. So, too, is an arc welder and a Zener diode. Typically, an arc needs a current limiting circuit to prevent a "runaway" current after the threshold voltage has been reached.

The damage that can be done by a spark or an arc depends entirely on the conditions. In a small circuit like a small neon (gas discharge) lamp, there is little damage than can be done. But a large sodium (gas discharge) lamp operates at a voltage and temperature that can cause serious injury. The damage that can be dome by an arc welder is obvious. Danger from sparks covers a wide range:

(a) Most humans can touch both terminals of a car battery and not feel anything. But if you accidentally short those terminals with a screw driver, the spark can be dangerous and can melt the end of the screw driver. The difference is the amount of current available. A voltage of 12 volts can't drive enough current through the resistance of a normal human body to cause injury. But the short circuit current can be 80 Amps or more, which can generate tremendous heat at the short.
(b) If you scuff your feet on a hotel carpet and then touch a door knob, you might feel a sharp (but not too dangerous) static electricity spark jump. In that situation the voltage is very high (say, 1,000 to 7,000 volts) but the current is negligible. If you hold the door key tightly in your fingers you likely won't feel the static spark. That is because the conducting area changes from a point on your finger to the area over which the key is held.
(c) If you are unfortunate enough to be hit by lightning……. well *both* the voltage and the current are high. A typical lightning flash is 300 million volts and 30,000 Amps!

5.2.1 Avoiding Shocks and Arcs: Sparks and arcs are initiated by large voltage differences. To avoid them, avoid large voltage differences! This might be accomplished by controlled circuits, by shielding, or by diode shunts for example. You can find circuits on the web for protecting electronic circuits from all but direct lightning strikes.

Notes:
1. Guglielmo Marconi's early wireless transmissions were initiated by very large spark jumps. A very good book telling a good story while also offering great insight into Marconi's early work is Erik Larson, "Thunderstruck," Crown Publishers, 2007.
2. See Coulomb's Laws in Section 1.2.
3. A dielectric is any material, including a gas, air, or vacuum. Every dielectric has its own physical properties.
4. Taken from or adopted from https://www.allaboutcircuits.com/textbook/direct-current/chpt-12/insulator-breakdown-voltage/

6. Physics of Trace Temperature

6.1 Joule Heating:

We already know that conductors heat by power dissipation in the conductor. The power dissipated in the conductor is related to the current (I) flowing in the conductor by the equation $P = I^2R$. This is known as Joule heating. In its most basic form, heat is motion (for example, at absolute zero temperature all motion stops.) When current flows down a trace, electrons move (Chapter 1). When current increases, electron motion increases. It is the effect of this motion that manifests itself as heat/temperature.

We intuitively know this can't be the full story, because if it was, the conductor would heat continuously. Conductors (usually) don't heat continuously; they usually reach a stable temperature and hold there until something (usually current) changes. At the same time a trace is being heated, it is being cooled by the *physics* of the materials surrounding the trace. A stable temperature is reached when the rate of heating equals the rate of cooling.

6.2 The Physics of Cooling:

PCB traces generally cool by three mechanisms, conduction, convection, and radiation.

6.2.1 Radiation: Thermal radiation is the emission of electromagnetic waves from all matter that has a temperature greater than absolute zero. Thermal radiation reflects the conversion of thermal energy into electromagnetic energy. The thermal energy is the kinetic energy of random movements of the electrons making up the current flow. Electromagnetic radiation, including visible light, will propagate indefinitely in vacuum (Note 1).

The effectiveness of thermal radiation from a trace depends on many variables that are beyond the scope of this book. But one of the most important physical properties related to radiation is emissivity (both of the radiating and the receiving materials.) One of the determinants of emissivity is color. Black objects radiate better than white objects, so changing the color associated with our circuits may be one way to help with the cooling process.

6.2.2 Convection: Convection, for our purposes, is the dissipation of heat from an object (pad or trace) through the air. In general, it is marginally efficient at cooling a heated object (component or trace) unless supplemental means are used. For example, if we heat a trace in laboratory conditions, the air immediately above the trace will heat, transferring some of the heat from the trace to the air above it. But that air may sit stagnantly above the trace, slowing any additional cooling. But if we incorporate supplemental cooling mechanisms, such as a fan, the heat transfer (and therefore the cooling) can be significantly improved.

There is a quantitative measure of radiation and convection cooling called the *Heat Transfer Coefficient*, HTC (Note 2). HTC applies to a wide range of physical situations; in our context it would apply to thermal convection and radiation. Within the field of thermodynamics there is much written about HTC and how to quantify it, but not much is written about how it applies to PCB traces. So, we can understand that increasing HTC results in cooler traces, but it is difficult to go much further.

6.2.3 Conduction: The primary cooling mechanism of PCB traces is conduction through the board material. Our interest would be in how heat is conducted through the dielectrics we use in our boards. Dielectrics have a property called the *Thermal Conductivity Coefficient*. On a data sheet it might be denoted by Tcon or k. Thermal conductivity applies in two different directions, typically called "in plane" or "through plane". The distinction is "in plane" (Tcon-x or Tcon-xy) is a measure of thermal conductivity in the same plane as the trace (horizontally, or xy-axis, across the board,) while "through plane" (Tcon-z) would be a measure of thermal conductivity vertically through the board in the z-axis. In-plane conductivity is typically better (higher) than through plane. The units of thermal conductivity are W/mK, or Watts per meter per degree Kelvin.

Thermal conductivity through a board material has a time constant associated with it (Note 3 and Section 6.5).

In our studies we found that dielectric thermal conductivity was one of the most significant determinants of trace temperature (after current.) The higher the thermal conductivity coefficient, the better the heat conduction away from the trace and the lower the trace temperature. While we can estimate trace temperatures through such studies as found in IPC 2152 (Note 4), typical variations in thermal conductivity between dielectrics can result in temperature differences of 10s of degrees.

6.3 Temperature is a Point Concept:

Increasing temperature is caused by the I^2R dissipation along a trace. Resistance (R) is a function of resistivity and of trace cross-sectional area, width x thickness. Both resistivity and thickness can vary from point to point along a trace and around the board. Therefore, the I^2R power dissipation varies from point to point. Therefore, the temperature can, and does, vary from point to point along a trace (Note 5).

6.4 Some Examples of the Physics of Cooling:

6.4.1 Board Parameters: When we purchase boards from a fabricator, we assume they are uniform. But they often are not. There might be a variation in thickness of the copper traces from one area of the board to another. This can result in significant thermal variations of otherwise "identical" circuits from one part of a board to another.

Processes might vary across board areas resulting in variations in trace thickness along a particular trace. This results in temperature variations from point to point along a trace (See again, Note 5.)

Often fabricators are not diligent in making sure parameters like thermal conductivity coefficient are uniform from one lot to the next, causing unexpected thermal variations from one lot to the next.

6.4.2 Copper Cored Board: If we place a conducting material, like a copper core between layers of a board or heat conducting ceramic particles imbedded in the dielectric, we can change the thermal conducting characteristics of the board and therefore the temperature profiles of heated pads/traces.

6.4.3 Heatsinks and Fans: A very common cooling strategy can be seen with computer processors. On some motherboards they run so hot that without supplemental cooling they would fail/melt in a very short time. It is common to see (1) large heatsinks placed over processors with (2) attached fans circulating air around them. This is an example of supplementing (the *physics* of) both the conduction and convection capabilities of the system.

6.4.4 Why Via Temperature Does Not Depend on Current: If we have a high-current carrying (power) trace connected to a trace on a different layer through a connecting via, we do not need to size the via to the trace. A much smaller via will do. This is because the parent trace acts as a heat sink and the *physics* of thermal conduction dominates the situation. The via, no matter what size it is (almost), cannot get much hotter than the trace because of the heat sinking capabilities of the trace (Note 6).

6.4.5 Boards Heat Through: Assume we have a heated pad or trace on the top layer of the board. We have previously shown that the bottom layer of the board will (in most cases) be with 10 °C of the top layer (Note 7). This is because of the thermal conducting properties through the board. For example, consider a typical 60 mil thick FR4 board 5" x 10" in size. There are two 1" square, 1.0 Oz. thick, pads on the top layer, one on each side. On the bottom layer there is a copper plane covering the left-hand side of the board. There is an 80 mil trace running across the board on a middle layer. The pads are heated by a 2.5 Watt source.

The thermal profile of the top layer is shown in Figure 6.1. The right-hand pad (without an underlying plane) is heated to 87 °C at its midpoint. The left-hand pad, with an underlying plane, is cooler, at 56 °C, as expected. Both pads illustrate how pads are cooler at their edges, where the thermal conduction is more efficient, and especially at their corners, where the cooling is most efficient.

We would expect the copper plane to draw heat away from the left-hand pad, and it does so. Since the plane is in contact with the 20 °C ambient, we might expect the plane's temperature to be near the ambient temperature. But the *physics* of thermal conduction through the board dominates here. Figure 6.2 illustrates the thermal image of the bottom layer. The peak temperatures underneath the heated pads are 49 °C and 84 °C, respectively. Even though a copper plane conducts heat very well, there is still a very significant *thermal gradient on the plane* underneath the pad. The temperature difference between the top and bottom layers under the pads is only 7 °C for the pad with an underlying plane and 3 °C for the pad without a plane!

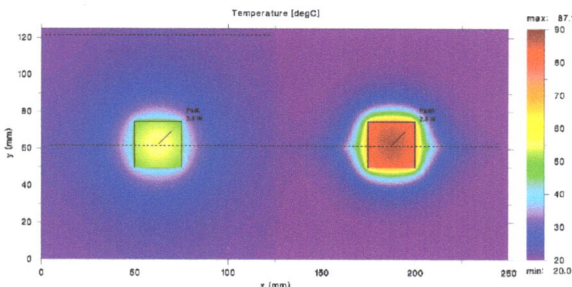

Figure 6.1
Thermal profile of top layer

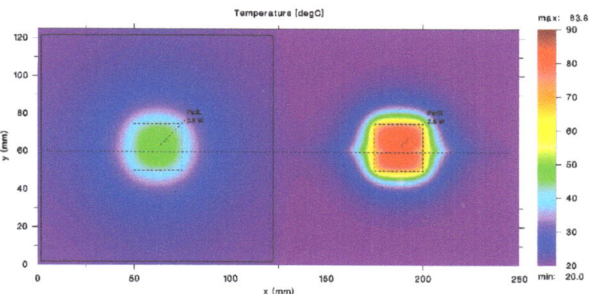

Figure 6.2
Thermal image of the bottom layer.

Also, all the internal layers inside the board, underneath the pads, are between these limits. For example, if we route a trace on an inner layer, even with no current through it at all, it will have a thermal profile similar to (in this case) the top layer. Figure 6.3 illustrates the thermal image of an 80 mil wide trace passing underneath the two pads. The thermal profile of the trace, even carrying no current itself, resembles the top layer profile.

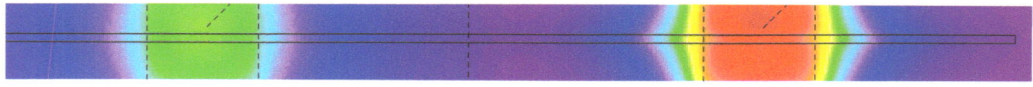

Figure 6.3
Thermal image of an inner layer showing it trace profile.

6.4.6 Why Thermal Vias are VERY Inefficient: Section 6.4.5, above, explains why thermal vias are *very* inefficient. If we drop a thermal via from the heated pad on the top layer, it has to terminate somewhere. The plane on the bottom layer is only a few (7) degrees cooler than the top layer. Thermal conductivity (through the thermal via) is a function of the temperature difference. If the temperature difference (between the heated pad and the bottom-layer plane) is small (as in this case), thermal conductivity is small. Therefore, each thermal via contributes only a small increment to the total cooling of the pad (again, see Note 7).

6.4.7 When Traces Melt: Figure 6.4 illustrates two different examples of traces melting under high current. Figure 6.4(a) illustrates the case of a sudden *fusing* of the trace due to a very high current overload. The trace fused (melted) in less than a second and the entire melting phase was over in a single camera frame (1/30 second.) Figure 6.4(b) shows the case of a marginal overload and the entire process took over 15 minutes. Note that the board damage was considerably worse for the marginal overload case.

(a)   (b)

Figure 6.4
Showing two different examples of traces melting

I. M. Onderdonk is credited with an equation for the current required to melt a trace (Note 8). Onderdonk's Equation is valuable because it includes the time required to meet the melting temperature. We can use the IPC 2152 data to estimate the temperature rise of a 20 mil wide trace carrying 3 Amps. And we can use Onderdonk's Equation to determine the fusing time for the same conditions. The results are shown in Figure 6.5 (Note 9).

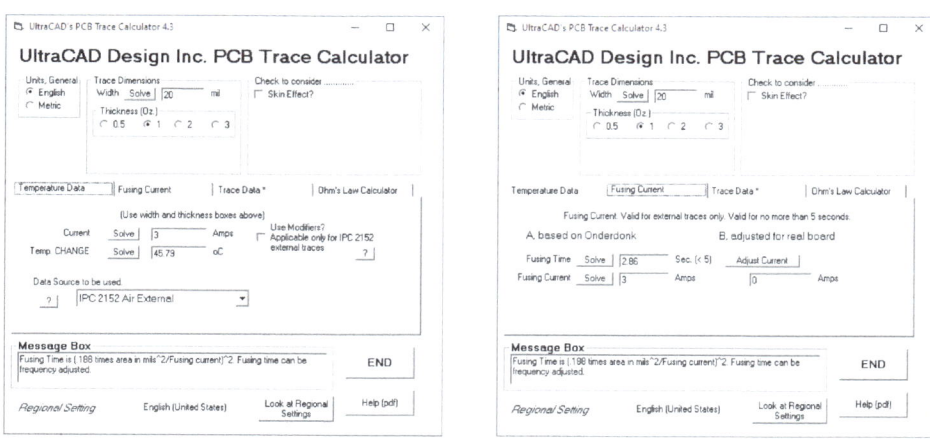

Figure 6.5
Temperature data for a 20 mil trace using UltraCAD's free PCB Trace Calculator.

IPC data shows the trace will reach a stable temperature of almost 66 °C (20 °C ambient plus a change of almost 46 °C.) But Onderdonk's Equation suggests the trace will reach 1083 °C and melt in less than 3 seconds! The difference is that Onderdonk's Equation *explicitly* assumes there is *no cooling* of the trace. The IPC data is based on an environment where cooling does exist. This illustrates the significant effect that *physics* (physical cooling in this case) can have on calculations and measurements.

(*Technical note*: Not much is known about Onderdonk, but he/she might have been involved with the electrical power distribution industry. The equation seems to have been developed for determining the fusing current for electrical wires in air, with an assumed reliability up to about 10 seconds, a time period during which the "no cooling" assumption might be reasonable for a wire in air.)

6.4.8 Thermal Time Constant: When current is applied to a trace, the trace starts to heat up. At first, it heats fairly quickly. But as the trace heats up, heat is conducted away from the trace, warming the material nearby, thereby slowing the rate at which the trace continues to warm up. The trace temperature typically has a time constant associated with it which depends on the heat transfer coefficient of the dielectric and the board geometry. But the shape of the heating curve is pretty typical. A representative one is shown in Figure 6.6 (Note 10). It typically takes from 5 to 15 minutes for normal trace temperatures to stabilize.

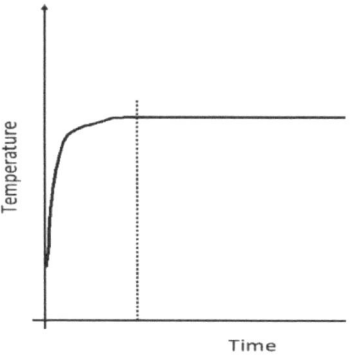

Figure 6.6
Trace temperature as a function of time.

Notes:
1. See https://en.wikipedia.org/wiki/Thermal_radiation
2. See https://www.thermopedia.com/content/841/
3. "PCB Design Guide to Via and Trace Currents and Temperatures," Brooks and Adam, Artech House, 2021.
4. IPC-2152, "Standard for Determining Current Carrying Capacity in Printed Board Design," August, 2009, www.IPC.org.
5. See Note 3, Chapter 13.
6. See Note 3, Chapter 8.
7. See Brooks and Adam, "PCB Signal Traces Are Hotter Than We Think", https://www.signalintegrityjournal.com/articles/2677-pcb-signal-traces-are-hotter-than-we-think
8. See Note 3. Section 11.4 talks about Onderdonk and his equation, Chapter 12 provides an extensive discussion of trace fusing and whether the time to fuse can be predicted. Dr. Adam provides a rigorous derivation of Onderdonk's Equation in Appendix G.
9. This calculator is part of UltraCAD's free Universal Calculator Package, See www.ultracad.com.
10. See Note 3, Section 7.2

Technical note: There are other material physical parameters that are temperature dependent, such as Glass Transition Temperature (Tg), Decomposition Temperature (Td), and Time to Delamination (T260/T288). But these parameters are typically not relevant at normal operating temperatures, so they are not discussed here.

7. Some Material Impacts

Of the several other dielectric material physical properties that we might discuss, three are particularly relevant for PCB designers: permittivity, permeability, and dielectric structure.

7.1 Permittivity:

Why are some capacitors able to store much more charge than others? What is the relative dielectric coefficient associated with dielectrics and why does it matter? The answer relates to a property called *permittivity*. Before we get too far into this topic, we need to clarify some terms.

Dielectric constant is an out-of-date term that means the same thing as permittivity, ε_m. The permittivity of a vacuum, ε_0, is given as 8.85418782 x 10^{12} Farads/meter. The relative dielectric constant, ε_r, or relative permittivity, is the ratio of the material permittivity to the permittivity in a vacuum, or

$$\varepsilon_r = \varepsilon_m / \varepsilon_0$$

ε_r is the variable we are familiar with that impacts such things as controlled impedance and signal propagation speed. *It is a measure of the electrical polarization that can occur in a material in response to an applied charge, or voltage, across the material*. Figure 7.1 illustrates a board with a differential pair of traces on either surface (or, it could be a trace over a plane.) The waveform illustrates a snapshot of a signal propagating along the trace. Prior to any signal, all electrical charges are distributed randomly throughout the material. When the signal polarity is one direction, the atoms and molecules in the dielectric are pulled in one direction (opposite charges attract, ref Coulomb in Chapter 1.) As the signal changes polarity, the atoms and molecules are pulled in the opposite direction. This action tends to electrically *polarize* the material.

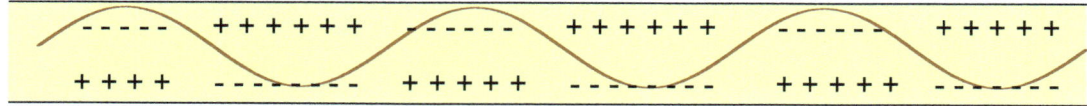

Figure 7.1
Signal propagating through a dielectric.

Materials with high relative permittivity polarize to a greater extent. That means they can store more charge. If such material is between the plates of a capacitor, that means the capacitance increases. Appendix 1 (to this chapter) provides relative permittivity values for several materials. Materials typically used in capacitors have relatively high relative permittivities (tantalum ≈ 25, ceramics ≈ 38.)

The *physics* of realigning the charges in a dielectric as a signal passes by takes energy. This energy is lost to the signal as it passes by. In one form the losses manifest themselves as dielectric losses. Dielectric losses are a function of frequency (they increase as the frequency increases) and of temperature. One practical consequence of dielectric losses is that the characteristic impedance of the trace changes (see Section 2.3.3.) This means there is no longer a single resistor value that can effectively terminate the trace. And this means that signal reflections may become an issue (Note 1).

Higher permittivities in dielectrics increase the capacitance between the trace and the underlying planes. Therefore, dielectrics with higher permittivities will lead to lower characteristic impedances for traces, affecting controlled impedance calculations. And since realigning charges absorbs energy from the signal, the signals will slow down in dielectrics with higher permittivities.

Note that permittivity relates to charged particles (electrons) and their inherent *charge*. This will contrast with permeability, below.

7.2 Permeability:

In electromagnetism, **permeability** is the *measure of magnetization that a material obtains in response to an applied magnetic field*. (Compare with *permittivity*, above.) Permeability is typically represented by the (italicized) Greek letter μ. In the macroscopic formulation of electromagnetism, there appears two different kinds of magnetic field:

- the magnetizing field **H** which is generated around electric currents and also emanates from the poles of magnets. The SI units of **H** are amperes/meter.
- the magnetic flux density **B** which acts **back** on the electrical domain, by curving the motion of charges and causing electromagnetic induction (Note 2). The SI units of **B** are volt-seconds/square meter (teslas).

Relative permeability, denoted by the symbol u_T, is the ratio of the permeability of a specific medium to the permeability of free space:
$$\mu_0: u_T = u/u_0$$
where $u_0 = 4\pi \times 10^{-7}$ H/m is the magnetic permeability of free space.

The importance of permeability is that it "amplifies" the magnetic field created by a current. For example, if we form a current-carrying conductor into a coil, as shown in Figure 7.2, there will be a magnetic (H) field generated inside the conductor loops. This will cause inductance in the conductor, the magnitude of which depends on the geometry of, and number of, the turns. If the core material inside the loops is air, the inductance will be relatively small. But if the core material has a high permeability (say an iron core) the magnetic field will be much larger, generating a much larger inductance in the coil. So, from an *electronic* standpoint, the magnitude of the inductance depends on the geometry of the coil. But the *physics* of the material inside that loop can have a dominating effect on that inductance.

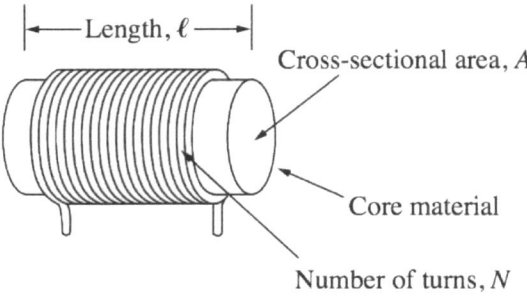

Figure 7.2
Wire wrapped around a toroid.

In terms of permeability, there are three types of materials, diamagnetic, paramagnetic, and ferromagnetic. Ferromagnetic materials are the only ones that significantly focus the magnetic field around a conductor. Circuit designers generally are only interested in ferromagnetic materials. Only four elements are ferromagnetic, Fe (iron), Co (cobalt), Ni (nickel), and Gd (gadolinium, look it up!)

The root cause of permeability is electron *spin* (as opposed to electron *charge* creating permittivity.) A current through a wire generates a magnetic field around the wire (Faraday's Law, Chapter 1). Electron *spin* creates a magnetic moment at the atomic level. In general, electron spin is chaotic and random for most materials. But the tendency (ability) of many of the electrons to align their spin in the same orientation is the characteristic that creates magnetic permeability. In some materials this orientation is only temporary and dissipates when the external magnetic field is removed. But with certain materials, and under certain conditions, a material can become permanently magnetized.

7.3 What Makes a Good Thermal Conductor

Remember from Chapter 6, in its most basic form, heat is motion. (For example, at absolute zero all motion stops.) At the atomic level, this motion includes atoms and especially their electrons. Increased heat (temperature) is related to the increased motion of electrons. Now if electrons are free to carry electrical energy through a metal, they are also free to carry heat energy. That's why metals that conduct electricity well are also good conductors of heat. Electrical insulators are also (generally) poor conductors of heat. (Things aren't quite so simple for nonmetals, however, because heat travels through them in other, more complex ways.)

7.4 Fiber Weave Effect:

Sometimes the specific way a material is physically fabricated can impact its electrical performance. For example, a typical PC board material is made up of resin impregnated fiberglass (Figure 7.3.) Note that in Figure 7.3(a) the fiberglass weave is very loose, while in 7.3(b) it is very tight. Resin and glass have significantly different permittivities (glass ≈ 6, resin ≈ 3.8). The ε_r of the board material will be specified as the combined average of the two (somewhere around 4.2 to 4.5 or so.) So, the board represented in 7.3(b) will have a higher permittivity (since it has more glass) than will 7.3(a).

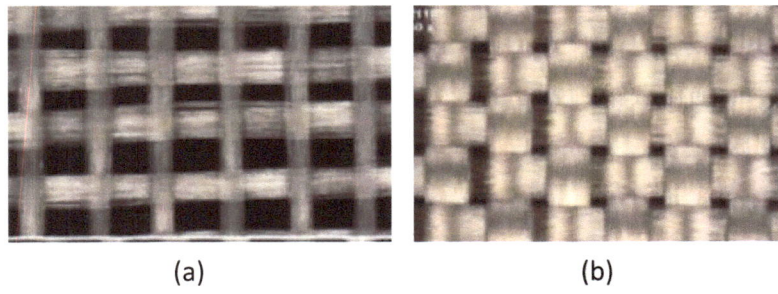

Figure 7.3
A typical board is fabricated from resin impregnated fiberglass.

That weave may also impact signal timing. We noted in Section 3.4 that signal propagation speed is the speed of light divided by the relative dielectric constant (relative permittivity, ε_r). Assume Figure 7.4(a) and (b) illustrate a differential pair routed across the boards shown in Figure 7.3. The traces in Figure 7.4(a) are routed in different environments (one over almost all glass, the other routed over some glass and some resin.) In Figure 7.4(b) they are both routed over almost all glass. In Figure 7.4(b) the traces will have the same propagation speed, but in Figure 7.4(a) the trace routed over all glass will be slightly slower than the other trace. This may cause an offset between the two differential signals which may result in an EMI problem.

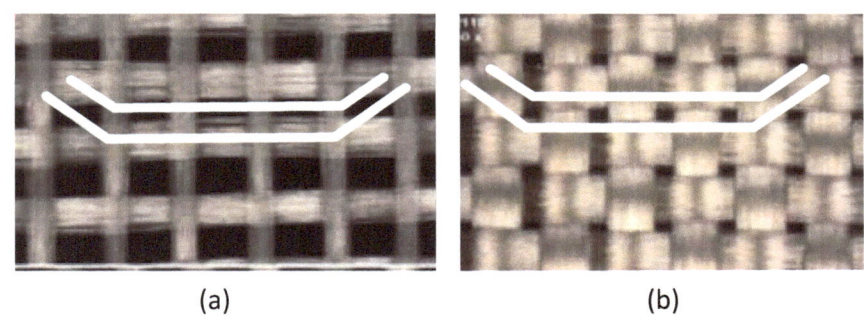

Figure 7.4
Traces routed over board shown in Figure 7.3 (a)

Notes:
1. See "PCB Currents; How They Flow, How They React," Doug Brooks, Prentice Hall, 2013, Chapter 20.
2. When we apply a changing current through a wire, the magnetic field it creates is the H field. That field creates a current in the opposite direction (causing inductance and skin effect, Section 2.3). The induced magnetic field caused by the new induced current in the backwards direction is the B field.

Appendix to Chapter 7
Relative Permittivity of Selected Materials

Material	Min.	Max.
Air	1	1
Amber	2.6	2.7
Asbestos fiber	3.1	4.8
Bakelite	5	22
Barium Titanate	100	1250
Beeswax	2.4	2.8
Cambric	4	4
Carbon Tetrachloride	2.17	2.17
Celluloid	4	4
Cellulose Acetate	2.9	4.5
Durite	4.7	5.1
Ebonite	2.7	2.7
Epoxy Resin	3.4	3.7
Ethyl Alcohol	6.5	25
Fiber	5	5
Formica	3.6	6
Glass	3.8	14.5
Glass Pyrex	4.6	5
Gutta Percha	2.4	2.6
Isolantite	6.1	6.1
Kevlar	3.5	4.5
Lucite	2.5	2.5
Mica	4	9
Micarta	3.2	5.5
Mycalex	7.3	9.3
Neoprene	4	6.7

Material	Min.	Max.
Nylon	3.4	22.4
Paper	1.5	3
Paraffin	2	3
Plexiglass	2.6	3.5
Polycarbonate	2.9	3.2
Polyethylene	2.5	2.5
Polyimide	3.4	3.5
Polystyrene	2.4	3
Porcelain	5	6.5
Quartz	5	5
Rubber	2	4
Ruby Mica	5.4	5.4
Selenium	6	6
Shellac	2.9	3.9
Silicone	3.2	4.7
Slate	7	7
Soil dry	2.4	2.9
Steatite	5.2	6.3
Styrofoam	1.03	1.03
Teflon	2.1	2.1
Titanium Dioxide	100	100
Vaseline	2.16	2.16
Vinylite	2.7	7.5
Water distilled	34	78
Waxes, Mineral	2.2	2.3
Wood dry	1.4	2.9

Source: https://sites.google.com/site/atspring2013/refraction---2009/relative-permittivity---beeswax-teflon-interface

Appendix 1

Some Basic Waveform Characteristics and Examples (Note 1)

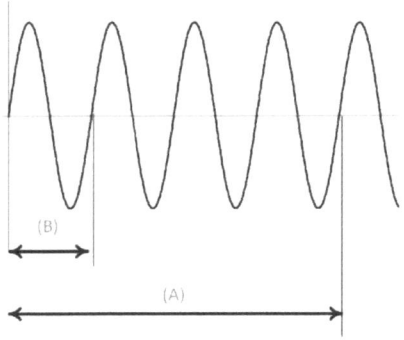

(Sine wave)
Example: Let (A) = one second of time.
    Frequency = number of cycles the waveform makes in one second of time.
        Frequency, f, = 4 cycles per second
        Frequency, ω, = 2πf = 25.1 radians per second
    Period (B) = length of time for one cycle = 1/f
        Period = ¼ second

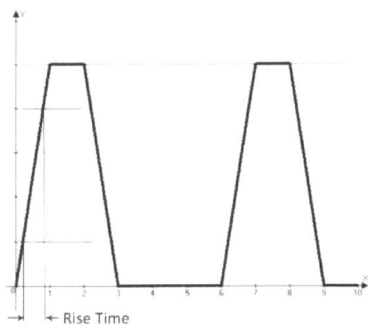

(Pulsed Waveform)
Rise/Fall Time = time between 20% and 80% signal levels
Duty Cycle = percentage of time signal is high
    = 1/3 = 33% in this example

Propagation Speed/Time:
    Speed: Signal velocity ≈ 12"/ns in vacuum
                  ≈ 6"/ns in FR4
    Time: Propagation time = 1/Propagation Speed
                  ≈ 1.0 ns/ft in vacuum
                  ≈ 2.0 ns/ft in FR4

Wavelength (λ) = *distance* between identical points on a waveform traveling along a conductor or in a space. Wavelength depends on propagation speed/velocity.
    λ = propagation speed * period = propagation speed / f

Other Waveforms:

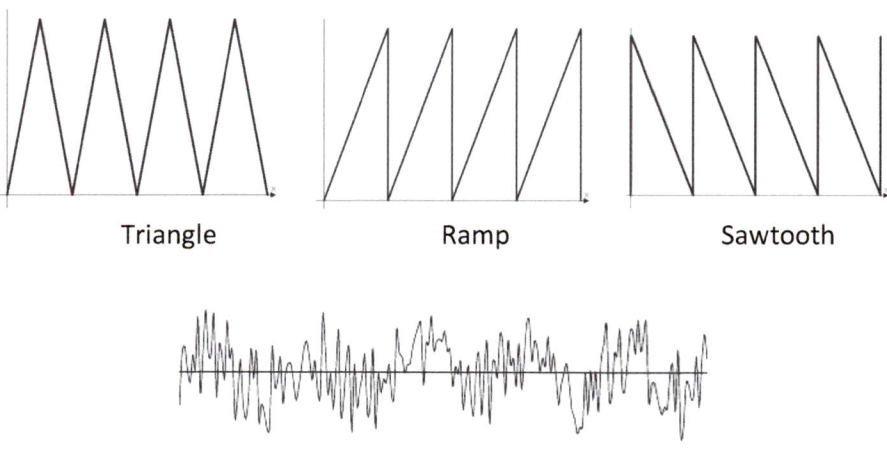

Triangle   Ramp   Sawtooth

Random

Fourier Series

Fourier's Theorem (paraphrased): *Every signal or curve, no matter what its nature may be, or in what way it was originally obtained, can be exactly reproduced by superposing a sufficient number of individual sine (and or cosine) waveforms of different frequencies (harmonics) and different phase shifts.*

Example: Square wave:

$$\text{Square}(\theta) = \cos(\theta) - \frac{\cos(3\theta)}{3} + \frac{\cos(5\theta)}{5} - \frac{\cos(7\theta)}{7} + \text{etc}$$

Free Calculator for illustrating Fourier Series (Note 2):

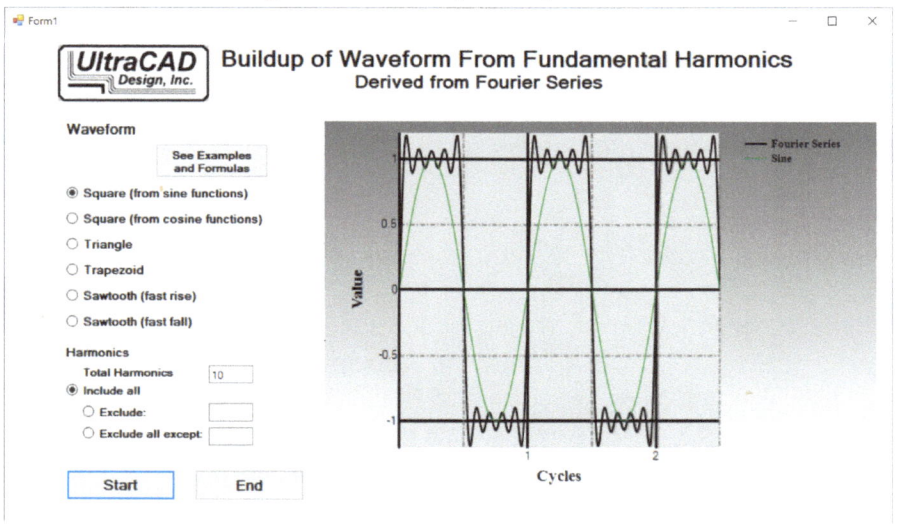

Types of Signal Modulation (Note 3)

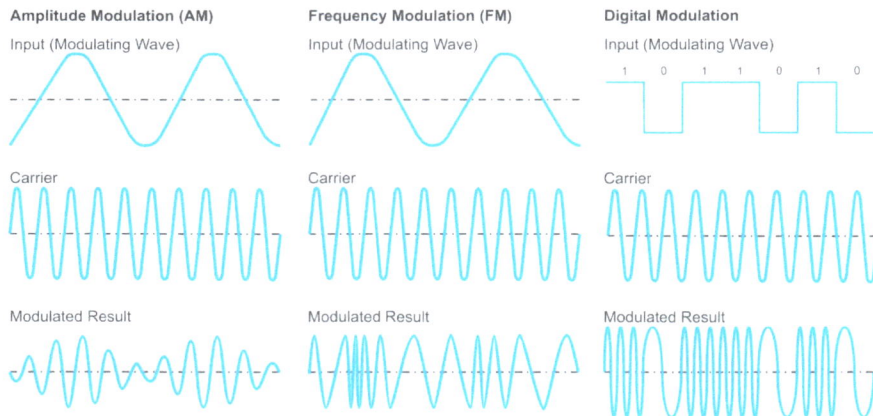

Notes:
1. For more information on this, see "PCB Currents; How They Flow, How They React," Doug Brooks, Prentice Hall, 2013, Chapter 2.
2. Source: Part of UltraCAD's free Universal Calculator package.
   http://www.ultracad.com
3. Source: https://www.taitradioacademy.com/topic/how-does-modulation-work-1-1/

Appendix 2

Some Definitions

Physics:
Physics is the natural science that studies matter, its fundamental constituents, its motion and behavior through space and time, and the related entities of energy and force. Physics is one of the most fundamental scientific disciplines, with its main goal being to understand how the universe behaves. A scientist who specializes in the field of physics is called a physicist.

(if used with a sing. verb) The science of matter and energy and of interactions between the two, grouped in traditional fields such as acoustics, optics, mechanics, thermodynamics, and electromagnetism, as well as in modern extensions including atomic and nuclear physics, cryogenics, solid-state physics, particle physics, and plasma physics.

Sources:
https://en.wikipedia.org/wiki/Physics

American Heritage® Dictionary of the English Language, Fifth Edition. Copyright © 2016 by Houghton Mifflin Harcourt Publishing Company. Published by Houghton Mifflin Harcourt Publishing Company. All rights reserved.

Electrical Engineering :
The branch of engineering that deals with the technology of electricity, especially the design and application of circuitry and equipment for power generation and distribution, machine control, and communications.

Source: American Heritage® Dictionary of the English Language, Fifth Edition. Copyright © 2016 by Houghton Mifflin Harcourt Publishing Company. Published by Houghton Mifflin Harcourt Publishing Company. All rights reserved.

Electronics:
Electronics is the study of electricity (the flow of electrons) and how to use that to build things like computers. It uses circuits that are made with parts called components and connecting wires to do useful things. The science behind Electronics comes from the study of physics and gets applied in real-life ways through the field of electrical engineering.

Many people can name several simple electronic components, such as transistors, fuses, circuit breakers, batteries, motors, transformers, LEDs and bulbs, but as the number of components starts to increase, it often helps to think in terms of smaller systems or blocks, which can be connected together to do something useful.

Source:
https://simple.wikipedia.org/wiki/Electronics

Other Books By the Author:

"PCB Design Guide to Via and Trace Currents and Temperatures," Brooks and Adam, Artech House, 2021.

"PCB Currents; How They Flow, How They React," Doug Brooks, Prentice Hall, 2013

"UltraCAD's Best Articles and Application Notes," Doug Brooks, 2022, available on Amazon.com

"Maxwell's Equations Without the Calculus," Douglas Brooks, 2016, available on Amazon.com

The Sampling Distribution and Central Limit Theorem, Douglas Brooks, 2012, available on Amazon.com

www.ingramcontent.com/pod-product-compliance
Lightning Source LLC
Chambersburg PA
CBHW051208220526
45473CB00003B/944